DUAL USE RESEARCH OF CONCERN IN THE LIFE SCIENCES

Current Issues and Controversies

Committee on Dual Use Research of Concern:
Options for Future Management

Committee on Science, Technology, and Law

Policy and Global Affairs

A Consensus Study Report of

The National Academies of
SCIENCES · ENGINEERING · MEDICINE

THE NATIONAL ACADEMIES PRESS
Washington, DC
www.nap.edu

THE NATIONAL ACADEMIES PRESS   500 Fifth Street, NW   Washington, DC 20001

This activity was supported with grants to the National Academy of Sciences from the Burroughs Wellcome Fund, the Gordon and Betty Moore Foundation, and the Alfred P. Sloan Foundation (#G-2014-13656) and by a contract between the National Academy of Sciences and the Federal Bureau of Investigation of the U.S. Department of Justice (#DJF-15-2300-P-0006686). Any opinions, findings, conclusions, or recommendations expressed in this publication do not necessarily reflect the views of any organization or agency that provided support for the project.

International Standard Book Number-13:   978-0-309-45888-7
International Standard Book Number-10:   0-309-45888-9
Library of Congress Control Number: 2017947966
Digital Object Identifier: https://doi.org/10.17226/24761

Additional copies of this publication are available for sale from the National Academies Press, 500 Fifth Street, NW, Keck 360, Washington, DC 20001; (800) 624-6242 or (202) 334-3313; http://www.nap.edu.

Copyright 2017 by the National Academy of Sciences. All rights reserved.

Printed in the United States of America

Suggested citation: National Academies of Sciences, Engineering, and Medicine. 2017. *Dual Use Research of Concern in the Life Sciences: Current Issues and Controversies.* Washington, DC: The National Academies Press. doi: https://doi.org/10.17226/24761.

The National Academies of
SCIENCES · ENGINEERING · MEDICINE

The **National Academy of Sciences** was established in 1863 by an Act of Congress, signed by President Lincoln, as a private, nongovernmental institution to advise the nation on issues related to science and technology. Members are elected by their peers for outstanding contributions to research. Dr. Marcia McNutt is president.

The **National Academy of Engineering** was established in 1964 under the charter of the National Academy of Sciences to bring the practices of engineering to advising the nation. Members are elected by their peers for extraordinary contributions to engineering. Dr. C. D. Mote, Jr., is president.

The **National Academy of Medicine** (formerly the Institute of Medicine) was established in 1970 under the charter of the National Academy of Sciences to advise the nation on medical and health issues. Members are elected by their peers for distinguished contributions to medicine and health. Dr. Victor J. Dzau is president.

The three Academies work together as the **National Academies of Sciences, Engineering, and Medicine** to provide independent, objective analysis and advice to the nation and conduct other activities to solve complex problems and inform public policy decisions. The National Academies also encourage education and research, recognize outstanding contributions to knowledge, and increase public understanding in matters of science, engineering, and medicine.

Learn more about the National Academies of Sciences, Engineering, and Medicine at **www.nationalacademies.org**.

The National Academies of
SCIENCES • ENGINEERING • MEDICINE

Consensus Study Reports published by the National Academies of Sciences, Engineering, and Medicine document the evidence-based consensus on the study's statement of task by an authoring committee of experts. Reports typically include findings, conclusions, and recommendations based on information gathered by the committee and the committee's deliberations. Each report has been subjected to a rigorous and independent peer-review process and it represents the position of the National Academies on the statement of task.

Proceedings published by the National Academies of Sciences, Engineering, and Medicine chronicle the presentations and discussions at a workshop, symposium, or other event convened by the National Academies. The statements and opinions contained in proceedings are those of the participants and are not endorsed by other participants, the planning committee, or the National Academies.

For information about other products and activities of the National Academies, please visit www.nationalacademies.org/about/whatwedo.

COMMITTEE ON DUAL USE RESEARCH OF CONCERN: OPTIONS FOR FUTURE MANAGEMENT

Co-Chairs

RICHARD A. MESERVE (NAE), Senior Of Counsel, Covington & Burling LLP
HAROLD E. VARMUS (NAS/NAM), Lewis Thomas University Professor, Weill Cornell Medicine

Members

ARTURO CASADEVALL (NAM), Professor and Chair, W. Harry Feinstone Department of Molecular Microbiology and Immunology, The Johns Hopkins Bloomberg School of Public Health
DENISE CHRYSLER, Director, The Network for Public Health Law–Mid-States Region, University of Michigan School of Public Health
ANUJ C. DESAI, William Voss-Bascom Professor of Law, University of Wisconsin
MICHAEL ETTENBERG (NAE), Managing Partner, Dolce Technologies
DAVID FIDLER, James Louis Calamaras Professor of Law, Indiana University Maurer School of Law
CLAIRE FRASER (NAM), Director of the Institute for Genome Sciences and Professor of Medicine, University of Maryland School of Medicine
MICHAEL HOPMEIER, President, Unconventional Concepts, Inc.
JAMES LE DUC, Professor, Department of Microbiology and Immunology and Director, Galveston National Laboratory, University of Texas Medical Branch, Galveston
W. IAN LIPKIN, John Snow Professor of Epidemiology, Professor of Neurology and Pathology, and Director of the Center for Infection and Immunity, Columbia University Mailman School of Public Health
STEPHEN S. MORSE, Professor of Epidemiology and Director, Infectious Disease Epidemiology Certificate Program, Columbia University Mailman School of Public Health

Staff

ANNE-MARIE MAZZA, Study Director and Senior Director, Committee on Science, Technology, and Law
JO L. HUSBANDS, Scholar/Senior Project Director, Board on Life Sciences
STEVEN KENDALL, Program Officer, Committee on Science, Technology, and Law
KAROLINA KONARZEWSKA, Program Coordinator, Committee on Science, Technology, and Law

KARIN MATCHETT, Consultant Writer
D. ALLEN AMMERMAN, Financial Officer, Committee on Science, Technology, and Law (until June 2017)
EILEEN N. ONI, Christine Mirzayan Science and Technology Policy Fellow

COMMITTEE ON SCIENCE, TECHNOLOGY, AND LAW

Co-Chairs

DAVID BALTIMORE (NAS/NAM), President Emeritus and Robert Andrews Millikan Professor of Biology, California Institute of Technology
DAVID S. TATEL, Judge, U.S. Court of Appeals for the District of Columbia Circuit

Members

THOMAS D. ALBRIGHT (NAS), Professor and Director, Vision Center Laboratory and Conrad T. Prebys Chair in Vision Research, Salk Institute for Biological Studies
ANN ARVIN (NAM), Vice Provost and Dean of Research, Lucile Salter Packard Professor of Pediatrics, and Professor of Microbiology and Immunology, Stanford University
CLAUDE R. CANIZARES (NAS), Bruno Rossi Professor of Physics, Massachusetts Institute of Technology
JOE S. CECIL, Project Director, Program on Scientific and Technical Evidence, Division of Research, Federal Judicial Center
R. ALTA CHARO (NAM), Warren P. Knowles Professor of Law and Bioethics, University of Wisconsin at Madison
HARRY T. EDWARDS, Judge, U.S. Court of Appeals for the District of Columbia Circuit
CHARLES ELACHI (NAE), Professor of Electrical Engineering and Planetary Science, Emeritus, California Institute of Technology
JEREMY FOGEL, Director, The Federal Judicial Center
HENRY T. GREELY, Deane F. and Kate Edelman Johnson Professor of Law and Professor, by courtesy, of Genetics, Stanford University
MICHAEL IMPERIALE, Arthur F. Thurnau Professor of Microbiology and Immunology, University of Michigan
ROBERT S. LANGER (NAS/NAE/NAM), David H. Koch Institute Professor, Massachusetts Institute of Technology
GOODWIN LIU, Associate Justice, California Supreme Court
JUDITH MILLER, Independent Consultant
JENNIFER MNOOKIN, Dean and David G. Price and Dallas P. Price Professor of Law, University of California, Los Angeles School of Law
DAVID A. RELMAN (NAM), Thomas C. and Joan M. Merigan Professor, Departments of Medicine, and of Microbiology and Immunology, Stanford University and Chief, Infectious Disease Section, VA Palo Alto Health Care System
MARTINE A. ROTHBLATT, Chairman and Chief Executive Officer, United Therapeutics

JOSHUA R. SANES (NAS), Professor of Molecular and Cellular Biology and Paul J. Finnegan Family Director, Center for Brain Science, Harvard University
WILLIAM B. SCHULTZ, Partner, Zuckerman Spaeder LLP
SUSAN S. SILBEY, Leon and Anne Goldberg Professor of Humanities, Professor of Sociology and Anthropology, and Professor of Behavioral and Policy Sciences, Massachusetts Institute of Technology
DAVID VLADECK, Professor and Co-Director, Institute for Public Representation, Georgetown Law School
SUSAN WESSLER (NAS), University of California President's Chair and Distinguished Professor of Genetics, University of California, Riverside
MICHELLE A. WILLIAMS (NAM), Dean of the Faculty, Harvard T.H. Chan School of Public Health

Staff

ANNE-MARIE MAZZA, Senior Director
STEVEN KENDALL, Program Officer
KAROLINA KONARZEWSKA, Program Coordinator
D. ALLEN AMMERMAN, Financial Officer (until June 2017)

Preface

In the 16 years since the 2001 anthrax letters mailings, the federal government, scientists, research institutions, and the international community have wrestled with the potential uses of peaceful biological research to do harm. Since 2001, there have been no public reports of serious biosecurity incidents in the United States. Nonetheless, concerns persist that a serious biosecurity event could occur, and there is a consistent desire to limit this possibility.[1] While numerous discussions[2] have taken place to consider approaches, policies, and mechanisms that would support an environment that optimizes the benefits of life sciences research while minimizing the possibility of the use of such research to do harm, consensus has not been reached domestically or internationally. Therefore, the following report examines the nation's policies on managing the dissemination of biological research of concern from conception to formal publication and offers findings that the committee hopes will inform future discussions and policies to manage such research.

---

[1] See, e.g., the reports and scorecards of the Blue Ribbon Study Panel on Biodefense. The study panel's website is http://www.biodefensestudy.org/index.htm.

[2] See, e.g., National Research Council, *Science and Security in a Post 9/11 World: A Report Based on Regional Discussions Between the Science and Security Communities* (Washington, DC: The National Academies Press, 2007), doi:https://doi.org/10.17226/12013; The Johns Hopkins Center for Health Security Conference, *Preserving National Security: The Growing Role of the Life Sciences*, March 3, 2011; Institute of Medicine and National Research Council, *Perspectives on Research with H5N1 Avian Influenza: Scientific Inquiry, Communication, Controversy: Summary of a Workshop* (Washington, DC: The National Academies Press, 2013), doi:https://doi.org/10.17226/18255; University of Maryland Center for Health and Homeland Security (CHHS) / Middle Atlantic Regional Center of Excellence for Biodefense and Emerging Infectious Diseases (MARCE) Conference, *Laboratory Safety, Security and Preparedness in the Evolving Era of Dual Use Research of Concern*, February 10, 2014; Institute of Medicine and National Research Council, *Potential Risks and Benefits of Gain-of-Function Research: Summary of a Workshop* (Washington, DC: The National Academies Press, 2015), doi:https://doi.org/10.17226/21666; National Academies of Sciences, Engineering, and Medicine, *Gain-of-Function Research: Summary of the Second Symposium, March 10-11, 2016* (Washington, DC: The National Academies Press, 2016), doi:https://doi.org/10.17226/23484.

We are indebted to the members of the committee who shared their experiences and knowledge with us, the experts who spoke to the committee, and the authors of papers commissioned to inform our deliberations. We also thank the staff at the National Academies of Sciences, Engineering, and Medicine—Anne-Marie Mazza, Jo Husbands, Steven Kendall, and Karolina Konarzewska—for their support in this project, and we express our thanks to consultant writer Karin Matchett.

Harold E. Varmus and Richard A. Meserve
Committee Co-Chairs

Acknowledgments

ACKNOWLEDGMENT OF PRESENTERS

The committee gratefully acknowledges the thoughtful contributions of the following individuals who made presentations before the committee at its July 2016 meeting:

Philip Campbell, *Nature*; Gerald L. Epstein, White House Office of Science and Technology Policy; Elisa D. Harris, University of Maryland; Teresa Hauguel, National Institute of Allergy and Infectious Diseases; Michael Imperiale, University of Michigan; Alan B. Morrison, The George Washington University; David A. Relman, Stanford University and VA Palo Alto Healthcare System; Randy Schekman, *eLife*; Ara Tahmassian, Harvard University; Inder Verma, *Proceedings of the National Academy of Sciences of the United States of America*; Carrie Wolinetz, National Institutes of Health; and David L. Wynes, Emory University.

The committee also would like to thank the authors of the commissioned papers who presented their work at the committee's January 2017 workshop. Their papers are available at https://www.nap.edu/catalog/24761 under the Resources tab. Those authors and the titles of their papers are:

Sam Weiss Evans, Harvard University—*The Construction of New Security Concerns in the Life Sciences*.
Michael Imperiale, University of Michigan and David A. Relman, Stanford University and VA Palo Alto Healthcare System—*Options for Management of Potentially Dangerous Information Generated by Life Science Research*.
Joseph Kanabrocki, The University of Chicago—*Biosafety, Biosecurity and Dual-Use Research of Concern*.

Duane Lindner and Winalee Carter,[1] Sandia National Laboratories—*Control of Sensitive Information: Policy, Procedure, and Practice in a National Security Context.*

Piers D. Millett, Biosecure Ltd.—*Gaps in the International Governance of Dual-Use Research of Concern.*

Tim Stearns, Stanford University—*Moving Beyond Dual Use Research of Concern Regulation to an Integrated Responsible Research Environment.*

Kimberly Strosnider, Doron Hindin, and Peter D. Trooboff,[2] Covington & Burling LLP—*The Role of Export Controls in Regulating Dual Use Research of Concern: Striking a Balance Between Freedom of Fundamental Research and National Security.*

Finally, the committee would like to thank the moderators of January 2017 workshop sessions:

Nancy Connell, Rutgers, The State University of New Jersey; Gigi Kwik Gronvall, The Johns Hopkins Bloomberg School of Public Health; and Margaret E. Kosal, Georgia Institute of Technology.

The commissioned papers and the presentations from both committee meetings are cited throughout this report.

ACKNOWLEDGMENT OF REVIEWERS

This Consensus Study Report was reviewed in draft form by individuals chosen for their diverse perspectives and technical expertise. The purpose of this independent review is to provide candid and critical comments that will assist the National Academies of Sciences, Engineering, and Medicine in making each published report as sound as possible and to ensure that it meets the institutional standards for quality, objectivity, evidence, and responsiveness to the study charge. The review comments and draft manuscript remain confidential to protect the integrity of the deliberative process.

We thank the following individuals for their review of this report: Philip Campbell, *Nature*; Susan Coller-Monarez, U.S. Department of Homeland Security; Diane Griffin, The Johns Hopkins Bloomberg School of Public Health; Gigi Kwik Gronvall, The Johns Hopkins Bloomberg School of Public Health; Gregory Koblentz, George Mason University; David Korn,

---

[1] Ms. Carter did not attend the January workshop. Dr. Lindner presented and discussed the paper with the committee at that event.

[2] Mr. Trooboff did not attend the January workshop. Ms. Strosnider and Mr. Hindin presented and discussed the paper with the committee at that event.

Harvard University; Marc Lipsitch, Harvard University; Peter Palese, Icahn School of Medicine at Mount Sinai; and Ara Tahmassian, Harvard University.

Although the reviewers listed above provided many constructive comments and suggestions, they were not asked to endorse the conclusions or recommendations of this report nor did they see the final draft before its release. The review of this report was overseen by Ann Arvin, Stanford University and Martin Philbert, University of Michigan. They were responsible for making certain that an independent examination of this report was carried out in accordance with the standards of the National Academies and that all review comments were carefully considered. Responsibility for the final content rests entirely with the authoring committee and the National Academies.

Contents

Summary		1
1	Introduction	9
2	The Current Policy Environment	23
3	Managing Dual Use Research of Concern	45
4	Committee Findings	73
Appendixes		
A	Biographical Information of Committee and Staff	81
B	Committee Meeting Agendas	91
C	Acronyms and Abbreviations	97

Boxes, Figure, and Tables

BOXES

2-1 Size of the Life Sciences Enterprise, 24
2-2 U.S. Government Policies Relevant to the Dissemination of Dual Use Research of Concern, 39

FIGURE

2-1 Comparison of the Scope of Different Policies for the Oversight of Life Sciences Research Involving Pathogens, 26

TABLES

1-1 Manuscripts Reviewed by the National Science Advisory Board for Biosecurity (NSABB), 16

2-1 The Initial and Current Charge of the National Science Advisory Board for Biosecurity (NSABB), 32

3-1 Principles for the Dissemination of Life Sciences Research of Concern, 67

Summary

The potential misuse of advances in life sciences research is raising concerns about national security threats. The current report examines the U.S. strategy for reducing biosecurity risks in life sciences research and considers mechanisms that would allow researchers to manage the dissemination of the results of research while mitigating the potential for harm to national security.

There is a growing tension between a scientific culture based on transparency and the need for secrecy to protect national security. While "most scientists would argue that the openness that characterizes much of the scientific research enterprise is the source of the extraordinary gains in scientific knowledge that have enriched us materially and intellectually,"[1] the ideal of a scientific culture based on principles of openness and transparency faces continuing challenges. One challenge relates to a concern that adversaries might take advantage of advances in science and technology for malicious purposes. This is particularly challenging in the biological sciences given recent dramatic advances, especially in the genetic engineering of pathogenic or potentially pathogenic micro-organisms, and fears that these advances could be exploited by non-state actors or terrorists. There is a recognition among some leaders in the scientific community of an informal social contract wherein "scientists as individuals and the international scientific community have a shared responsibility, together with other members of society, to do their utmost to assure that scientific discoveries are used solely to promote the common good."[2] This premise is not, however, accepted by all scientific practitioners.

---

[1] National Research Council, *Biotechnology Research in an Age of Terrorism* (Washington, DC: The National Academies Press, 2004), doi:https://doi.org/10.17226/10827, pp. 99-100.

[2] International Council for Science (ICSU), *Freedom, Responsibility, and Universality of Science* (Paris: International Council for Science, 2014), p. 5. Available at http://www.icsu.org/publications/cfrs/freedom-responsibility-and-universality-of-science-booklet-2014/CFRS-brochure-2014.pdf.

In today's world of rapidly advancing science, where tools and technologies are more widely available than ever before and where the dissemination of scientific findings occurs through multiple channels and at multiple levels, developing policies for managing the dissemination of knowledge, tools, and techniques produced by scientific research has become ever more difficult.

In view of ongoing concerns about the communication of biological research results that might present significant risks and in the wake of incidents (such as the 2001 anthrax mailings)[3] in which naturally occurring biological materials were used for nefarious purposes, the United States has given significant attention to policies and practices that can enhance biosecurity.

CHARGE TO THE COMMITTEE

Our committee was charged with reviewing policies associated with dual use research of concern (DURC).[4] Its objective was to review possible mechanisms for managing dissemination of research findings that strike an appropriate balance between the value of openness in scientific research and the needs of national security. As such, this encompasses the roles and responsibilities of students, researchers, institutions, and the federal government in the conduct of research. While one might think of dissemination in terms of publication, the committee, with encouragement from the project's sponsors, considered the management of dissemination as occurring along a spectrum from idea generation to the formal publication of research results in journals.[5]

The committee gathered information both at a public information gathering meeting on July 11-12, 2016, and at a public workshop on January 4, 2017. To assist in its deliberations, the committee commissioned papers on a range of topics including biosafety and biosecurity, international approaches to biosecurity, ethics, export controls, and current government policies on information control (these papers are available at https://www.nap.edu/catalog/24761 under the Resources tab). Authors were asked explicitly to consider the impli-

---

[3] Two sources of information about biosecurity incidents since 1900 are W. S. Carus, *Bioterrorism and Biocrimes: The Illicit Use of Biological Agents Since 1900* (Washington, DC: Center for Counterproliferation Research, National Defense University, 2001) and K. Berger et al., "Biosecurity Risk Assessment of Acts Targeting a Laboratory" in Gryphon Scientific, *Risk and Benefit Analysis of Gain of Function Research: Final Report–April 2016* (Takoma Park: Gryphon Scientific, 2016). Available at http://www.gryphonscientific.com/wp-content/uploads/2016/04/Risk-and-Benefit-Analysis-of-Gain-of-Function-Research-Final-Report.pdf.

[4] As discussed further herein, "Dual use research of concern" is a term of art that refers to research that involves a particular set of agents and toxins and type of experiment.

[5] Points along the spectrum include, for example, the point where research is funded, the period when research is being conducted, the transmission of information about research through informal communications among researchers, presentations at meetings and conferences, training and teaching, and the circulation of draft manuscripts and pre-prints or other self-published papers through traditional or electronic means.

cations of restrictions on the dissemination of scientific information, and it was expected that contrasting viewpoints would be expressed during the course of interaction between the committee, the audience, and session moderators. The authors of the commissioned papers presented their work at the January workshop and participated in a discussion with the committee and attendees.

The committee hopes that the current report and its findings will provide policymakers with information for further deliberation.

COMMITTEE FINDINGS

The committee considered expert presentations given before it, the content of commissioned papers and related external materials, and public discussions and engaged in private deliberations. It offers the following list of findings on the state of managing dissemination of DURC. It hopes that these findings provide a baseline for the development of principles that will, in turn, lay the framework for government policy for managing the dissemination of information about the conduct and results of DURC research by federal agencies, the research community, and the international scientific community. In alignment with its charge, the committee is not offering recommendations.

CONTEXT: CHANGES IN RESEARCH AND COMMUNICATION TOOLS

A confluence of factors—including advancing technologies and technical capabilities, globalization, rapid sharing of information, the changing nature of scientific publication, and the capacity and intent of some to cause harm—has led to concerns about the dissemination of scientific information that could be directly exploited for nefarious purposes.

Scientific information is disseminated through a wide range of means including education, training, presentations and posters at conferences, pre-print servers, informal communications, patents, and formal publication. The prevalence of digital information and online transmission and storage of information related to dual use research also makes information increasingly vulnerable to hacking. Much of current policy, however, tends to focus on formal publication.

There are some oversight mechanisms in place to make decisions about the publication of information that might pose risks to biosecurity. To date, the number of instances where detailed review has occurred and the frequency with which information has been restricted (by voluntary redaction, use of export controls, etc.) is small.

Findings

1. In general, the United States has a solid record with regard to the safe conduct of biological research. Given the lack of a comprehensive reporting system, knowledge of the nature and full extent of biosafety and biosecurity incidents is incomplete. Nevertheless, the number of documented, publicly known incidents of serious biosafety errors or lapses of biosecurity at laboratories has been small.[6]
2. In the wake of concerns that biological materials could be used for nefarious purposes and the significant risks that communication of the results of some biological research might convey, the United States has given significant attention to policies and practices that can enhance biosecurity.
3. Even with regard to research that could be directly applied to bioterrorism, there are concerns about excessive restrictions on the free flow of information. Open dissemination of research findings, a fundamental principle of research practice, can serve to alert relevant communities to a risk, provide the foundation for the development of countermeasures, and establish the foundation for scientific advances that could have significant public health benefits.

U.S. GOVERNMENT POLICY

Many policies potentially apply to the dissemination of DURC. U.S. DURC policies provide structures for managing the dissemination of information about certain pathogens and types of experiments that raise biosecurity concern, but they apply only to research that is conducted at institutions receiving federal funding. Non-compliance presents the potential risk of the withdrawal of federal funding, but it is not clear whether other sanctions would, in fact, be imposed.

Findings

4. The dissemination of life sciences information that may raise biosafety and biosecurity concerns is governed by fragmented policies and regulations.
5. Federal policies on DURC reach only a portion of the individuals conducting life sciences research. Those conducting research at institutions that do not receive federal funds (whether in private industry, in the "Do-It-Yourself" community, in other nations, etc.) are not bound

---

[6] The committee is not suggesting that errors and lapses are inconsequential, as it recognizes that a single lapse could have significant policy and public health consequences.

by these policies, but other regulations, such as export control laws, could apply.
6. Research that might be considered as DURC can, in principle, be identified before it is carried out or during the course of work when an unusual finding is encountered. Policies for identification of DURC in early phases, with consequent actions (a decision not to fund the research, withdrawal of funding, classification, mitigation plans, etc.), are in place for some types of research. Intervention at an early stage is more appropriate and likely to be more effective than at the time of publication.
7. The current policy focus and definition of DURC do not capture biosecurity concerns in all relevant areas of life sciences research, especially those that are emerging (e.g., synthetic and systems biology, computational modeling, genome editing, gene drives, neuroscience, the isolation of new micro-organisms and toxins). On the other hand, the current system of DURC policies and regulations may constrain certain types of research [e.g., research with select agents and toxins, research with pathogens of pandemic potential (PPPs)] more than is necessary to serve legitimate biosecurity goals.
8. When the government does not fund the research in question, the First Amendment imposes strong limits on the government's ability to restrict the communication of research results, including research that could be used for bioterrorism. When the government funds the research in question, the First Amendment gives it more leeway to restrict the communication of research results, but even in that context, the government's authority may be constrained.
9. Currently, no international organization is giving systematic attention to developing policy or guidance regarding the dissemination of scientific information of concern. Potential mechanisms and institutions [e.g., the World Health Organization (WHO), the Biological Weapons Convention (BWC), the Australia Group (AG), the United Nations Security Council (UNSC), etc.] exist that could fulfill this function. There has been a recent decline in policy activity at the international level despite the fact that there are ongoing concerns and discussions about specific technologies (e.g., CRISPR-Cas9).
10. Export controls do not limit communications among U.S. citizens within the United States. Export controls thus have a limited reach and do not offer a mechanism, in and of themselves, to control the dissemination of information.

MECHANISMS AND PROCESSES

A key issue identified during the committee's public meetings and private discussions was how to provide researchers—and particularly journal editors—with guidance about potentially problematic research findings or manuscripts. DURC policies provide mechanisms to guide those carrying out federally funded research or working at institutions that receive federal funds, including requirements to develop, in appropriate cases, risk mitigation plans. Other researchers and journal editors do not have ready access to such guidance. In light of the increasing number of journals in many parts of the world and the utilization of pre-print servers and other means of online publication prior to (or in lieu of) traditional peer review, the situation is significantly more complicated. The following findings relate to U.S. researchers and their international collaborators.

Findings

11. There is no systematic process through which journal editors and researchers outside federally funded institutions can seek guidance from U.S. government experts on the management of manuscripts or on research activities that raise potential biosecurity concerns.
12. There is no shared, consistent policy among U.S. and international journals for addressing DURC.
13. There are limited mechanisms [e.g., the National Science Advisory Board for Biosecurity (NSABB),[7] Federal Bureau of Investigation Weapons of Mass Destruction Directorate coordinators] for ongoing engagement between the scientific community and the national security and intelligence communities on biosecurity issues.
14. As a federal advisory body, the NSABB does not have the legal authority to restrict the dissemination of information. The NSABB may provide advice regarding the publication of information only under narrowly defined circumstances. Moreover, knowledge and use of the NSABB throughout the research community is limited.
15. In contrast to the Recombinant DNA Advisory Committee process, the oversight of DURC does not include mechanisms for assessing and sharing of best practices in the management of biosecurity among research institutions or opportunities for high-level review and consultation.
16. In principle, the NSABB could provide a mechanism to fulfill many of the functions described above, but its current mandate is limited.

---

[7] Information about the National Science Advisory Board for Biosecurity (NSABB) may be found at http://osp.od.nih.gov/office-biotechnology-activities/biosecurity/nsabb. The work of the NSABB is discussed further herein.

EDUCATION AND TRAINING

Reaching consensus on the management of DURC is complicated by the fact that experts have fundamentally divergent views about the nature of the biosecurity threat.[8] Any effort to place controls on information for biosecurity purposes involves a careful consideration of the nature of the research, the risks of malevolent uses of the research results, the benefits for scientific advance or the development of countermeasures through open communication, and evaluation of means to reap the benefits while limiting the risks. Effective assessment relies on an appropriate knowledge of risk and policy options among the international community of researchers, funders, and publishers.

Findings

17. Despite the attention given to periodic controversies over DURC, the available evidence suggests that most life scientists have little awareness of issues related to biosecurity. Those training to become life scientists are rarely introduced to the topic in a systematic way. Education and training programs at the undergraduate, graduate, and postdoctorate levels generally do not include courses or discussions about dual use research or DURC, unless the student or trainee is involved in research with a select agent. Even in this case, biosafety is the primary focus. This situation hampers efforts to implement policies to address potential biosecurity risks, particularly in emerging research fields that may pose concerns.
18. The management of the dissemination of scientific information requires local, national, and international approaches to provide awareness-raising, education and training, and ongoing guidance and opportunities to share best practices and develop common approaches.
19. There are some extensive and effective programs at research institutions that deal with specific pathogens that ensure that researchers are trained in biosafety, but they are not systematically in place across U.S. research institutions. In a number of cases, the scope of these programs includes biosecurity and enables these particular communities to develop sophisticated views about these issues. Expanding these programs beyond a focus solely on specific pathogens could increase the ability of the broader research community to take greater responsibility for safeguarding dangerous information in ways that do not impede scientific advances.

---

[8] See, e.g., C. Boddie et al., "Assessing the Bioweapons Threat," *Science*, August 21, 2015, Vol. 349, No. 6250, pp. 792-793.

20. Lessons learned from experiences with efforts to manage the dissemination of research information are not being adequately assessed or shared so as to promote more effective practice.
21. Many investments have been made by major donors to assist foreign countries with enhancements to their biosafety capacity. Investments have also been made in some aspects of biosecurity (e.g., physical security, access controls, pathogen accounting, etc.). Far fewer resources have been devoted to awareness-raising, education and training, and policy development related to the conduct of research and the dissemination of scientific information that could be employed for bioterrorism.

CONCLUSION

Despite decades of effort, there is little national or international consensus with regard to appropriate policies for addressing issues associated with the conduct and dissemination of life sciences research that might qualify as DURC. The absence of an international commitment to addressing such issues; the lack of agreement regarding a framework for assessing risk, uncertainty, and benefit; and the difficulties the U.S. government has faced in developing policies that effectively manage DURC illustrate the challenges of resolving the issues concerning information dissemination raised by DURC.

1

Introduction

The potential misuse of advances in life sciences research is raising concerns about national security threats. The current report examines the U.S. strategy for reducing biosecurity risks in life sciences research and considers mechanisms that would allow researchers to manage the dissemination of the results of research while mitigating the potential for harm to national security.[1] We begin this report by tracing the development of ideas about the dissemination of scientific information, broadly defined, in the United States.

There is a growing tension between a scientific culture based on transparency and the need for secrecy to protect national security. While "most scientists would argue that the openness that characterizes much of the scientific research enterprise is the source of the extraordinary gains in scientific knowledge that have enriched us materially and intellectually,"[2] the ideal of a scientific culture based on principles of openness and transparency faces continuing challenges. One challenge relates to a concern that adversaries might take advantage of advances in science and technology for malicious purposes. This is particularly challenging in the biological sciences given recent dramatic advances, especially in the genetic engineering of pathogenic or potentially pathogenic micro-organisms, and fears that these advances could be exploited by non-state actors or terrorists. There is a recognition among some leaders in the scientific community of an informal social contract wherein "scientists as individuals and the international scientific community have a shared respon-

---

[1] Numerous proposals for handling the dissemination of sensitive dual-use information have been suggested. See, e.g., R. A. Zilinskas and J. B. Tucker, "Limiting the Contribution of the Open Scientific Literature to the Biological Weapons Threat," *Journal of Homeland Security*, December 2002 and National Research Council, *Science and Security in a Post 9/11 World: A Report Based on Regional Discussions Between the Science and Security Communities* (Washington, DC: The National Academies Press, 2007), doi:https://doi.org/10.17226/12013.
[2] National Research Council, *Biotechnology Research in an Age of Terrorism* (Washington, DC: The National Academies Press, 2004), doi:https://doi.org/10.17226/10827, pp. 99-100.

sibility, together with other members of society, to do their utmost to assure that scientific discoveries are used solely to promote the common good."[3] This premise is not, however, accepted by all scientific practitioners.

In today's world of rapidly advancing science, where tools and technologies are more widely available than ever before and where the dissemination of scientific findings occurs through multiple channels and at multiple levels, developing policies for managing the dissemination of knowledge, tools, and techniques produced by scientific research has become ever more difficult.

The balance between minimizing the risks and maximizing the benefits of research requires consistent attention, as do the mechanisms for the oversight of such research. Any discussion of risk necessitates a consideration of uncertainty.[4] In this context, it is important to consider whether, among the broader scientific community, there is appropriate awareness of the issues and policies related to life sciences research with the potential for dual use and whether limits placed on research and dissemination are reasonable and serve both scientific and security interests.

BIOSAFETY AND BIOSECURITY

Risks from biological research can result from lapses in biosafety and biosecurity. Biosafety policies focus on ensuring that research practices prevent laboratory accidents from creating risks of exposure to infectious pathogens for researchers, laboratory workers, and the general public. Biosecurity is related to the procedures that are intended to keep information or materials from individuals or groups seeking to use such information or materials for malicious purposes. While concerns about security risks arising from communication about scientific research fall within the realm of biosecurity, strong biosafety practices promote responsible research practices that provide a foundation for many elements of effective biosecurity. Moreover, research that raises significant biosafety issues may also prompt concerns about potential biosecurity risks.[5]

---

[3] International Council for Science (ICSU), *Freedom, Responsibility, and Universality of Science* (Paris: International Council for Science, 2014), p. 5. Available at http://www.icsu.org/publications/cfrs/freedom-responsibility-and-universality-of-science-booklet-2014/CFRS-brochure-2014.pdf.

[4] Risk is the probability or threat of a negative occurrence: when conducting an assessment of risk, it is possible to accurately calculate the odds of a probable outcome. Uncertainty occurs where possible outcomes are known but probabilities cannot be attached to them. With any given problem (e.g., should a particular paper be published without redaction?), there is a sense of the outcome (or outcomes) to be avoided (e.g., an individual with nefarious intent using information about a particular pathogen to cause harm "to public health and safety, agricultural crops and other plants, animals, the environment, materiel, or national security"). What isn't known is the probability of a particular outcome.

[5] The controversy over the publication of papers describing the means to increase transmissibility of the H5N1 influenza virus that led to the creation of several key U.S. policies to oversee DURC is

INTRODUCTION

In general, the United States' record with regard to the safe conduct of biological research appears solid, but data are incomplete. There are cases where laboratory workers have died,[6] but the number of documented biosafety incidents that resulted in serious harm to human health has been very small. Similarly, only one serious biosecurity incident—the mailings of anthrax to members of Congress and the media in October 2001—has occurred in the United States. Nevertheless, given the size of the biological research enterprise, the diversity of research and research institutions, and the lack of a uniform reporting system, it is difficult to determine the full extent of lapses in biosafety.

---

the most recent example of these connections. The National Science Advisory Board on Biosecurity (NSABB) recommended against publication of the two gain-of-function (GOF) papers because they presented a threat to biosecurity and "argued that 'publishing these experiments in detail would provide information to some person, organization or government that would help them to develop similar mammal-adapted influenza A/H5N1 viruses for harmful purposes.'" However, "many national-security experts and scientists objected to the work simply because they believed it was not safe." See G. Kwik Gronvall, *H5N1: A Case Study for Dual-Use Research* (New York: Council on Foreign Relations, July 2013). By the time of the White House announcement in October 2014 of a pause in funding for certain GOF experiments, biosafety and biosecurity concerns were receiving equal weight: "Gain-of-function studies may entail biosafety and biosecurity risks; therefore, the risks and benefits of gain-of-function research must be evaluated, both in the context of recent U.S. biosafety incidents and to keep pace with new technological developments, in order to determine which types of studies should go forward and under what conditions [White House, *U.S. Government Gain-of-Function Deliberative Process and Research Funding Pause on Selected Gain-of-Function Research Involving Influenza, MERS, and SARS Viruses*, (Washington, DC, 2014a). Available at: http://www.phe.gov/s3/dualuse/Documents/gain-of-function.pdf].

[6] The introduction to the 5th edition of *Biosafety in Microbiological and Biomedical Laboratories* (BMBL) [U.S. Department of Health and Human Services, *Biosafety in Microbiological and Biomedical Laboratories*, 5th ed. (Washington, DC, 2009)] includes a discussion of the data available about laboratory acquired infections (LAIs) and provides the information available at the time of publication (2009) about fatalities. For example, the BMBL cites studies by Pike and Sulkin [see S. E. Sulkin and R. M. Pike, "Survey of Laboratory-acquired Infections," *American Journal of Publich Health*, 1951: Vol. 41, pp. 769-781; R. M. Pike, S. E. Sulkin, and M. L. Schulz "Continuing Importance of Laboratory-acquired Infections," *American Journal of Public Health*, 1965, Vol. 55, pp. 190-199; R. M. Pike, "Laboratory-associated Infections: Summary and Analysis of 3921 Cases, *Health Laboratory Science*, 1976, Vol. 13, pp. 105-114; R. M. Pike, "Past and Present Hazards of Working with Infectious Agents," *Archives of Pathology and Laboratory Medicine*, 1978, Vol. 102, pp. 333-336; and R. M. Pike, "Laboratory-Associated Infections: Incidence, Fatalities, Causes, and Prevention, *Annual Review of Microbiology*, 1979, Vol. 33, pp. 41-66] that identified 4,079 LAIs between 1930 and 1978 that resulted in 168 deaths. "During the 20 years following the Pike and Sulkin publications, a worldwide literature search by Harding and Byers [see A. L. Harding and K. B. Byers, "Epidemiology of Laboratory-associated Infections," in D. O. Fleming and D. L. Hunt, eds., *Biological Safety: Principles and Practices*, 3rd ed. (Washington, DC: ASM Press, 2000), pp. 35-54] revealed 1,267 overt infections with 22 deaths" (see BMBL, p. 2).

"DUAL USE" RESEARCH

In the United States, policies related to security concerns about scientific research have traditionally focused on research results that have both civilian and military applications. Such research has come to be known as "dual use" research.[7] Initially, emphasis was placed on certain kinds of research in the physical sciences and engineering, with nuclear physics as the classic example. During the Cold War, the United States and its NATO allies constructed national and international frameworks, including coordinated export control regimes, to prevent advances in Western science and technology from reaching the Soviet Union and its allies.

By the late 1990s, incidents such as the bombing at the World Trade Center and the Aum Shinrikyo attacks on the Tokyo subway raised the specter of "mass casualty terrorism," including through the use of biological agents.[8] The anthrax mailings in the wake of the September 11, 2001, attacks in New York City and Washington, DC, thrust bioterrorism into public awareness.

Earlier in 2001, concerns about biological research had been raised when researchers in Australia published the results of a study that led to the creation of a highly virulent strain of mousepox that was lethal even to mice that had been vaccinated for naturally occurring mousepox.[9] This publication was followed by a paper that investigated the basis for the difference between the virulence factors in variola major virus, which causes smallpox, and vaccinia virus, which is used as a vaccine against the disease.[10] While, to some, the research provided valuable information for those seeking to understand and treat infectious disease, to others these papers represented a type of open publication that could provide a "roadmap" for terrorists seeking to weaponize biological agents.[11]

---

[7] In this case, dual use also could have positive connotations, where investments in military research and development could lead to valuable civilian "spin-offs."

[8] J. C. Gannon, "Viewing Mass Destruction Through a Microscope," *New York Times,* Section E, p. 10, October 11, 2001, and D. Hearst, "Smart Bio-Weapons Are Now Possible," *The Guardian.* May 20, 2001, available at http://www.guardian.co.uk/uk_news/story/0,3604,959473,00.html. It should be noted that the international community was also concerned about the discovery in Iraq, after the Persian Gulf War, of efforts by Iraq to develop biological weapons and the discovery, after the fall of the Soviet Union, that the Soviets had continued their offensive biological weapons programs after joining the Biological Weapons Convention in the early 1970s.

[9] R. J. Jackson et al., "Expression of Mouse Interleukin-4 by a Recombinant Ectromelia Virus Suppresses Cytolytic Lymphocyte Responses and Overcomes Genetic Resistance to Mousepox," *Journal of Virology,* February 2001, Vol. 75, pp. 1205-1210.

[10] A. M. Rosengard et al., "Variola Virus Immune Evasion Design: Expression of a Highly Efficient Inhibitor of Human Complement," *Proceedings of the National Academy of Sciences of the United States of America,* June 25, 2002, Vol. 99, No. 13, pp. 8808-8813.

[11] G. L. Epstein "Controlling Biological Warfare Threats: Resolving Potential Tensions Among the Research Community, Industry, and the National Security Community," *Critical Reviews in Microbiology*, January 1, 2001, Vol. 27, No. 4, pp. 321-354.

In 2004, editors of major life sciences journals published a joint "Statement on Scientific Publication and Security." The editors affirmed that "there is information that, though we cannot capture it with lists or definitions, presents enough risk of use by terrorists that it should not be published." They continued by saying, "how and by what processes it might be identified will continue to challenge us, because . . . it is also true that open publication brings benefits not only to public health but also in efforts to combat terrorism." The journal editors: (1) affirmed peer-reviewed journals' responsibility to publish high-quality research in enough detail to permit reproduction of the experiments; (2) affirmed their commitment to dealing responsibly with safety and security issues that arise; (3) urged scientists and journals to develop processes to deal with papers that may pose security risks; and (4) affirmed that "on occasion an editor may conclude that the potential harm of publication outweighs the potential societal benefits" and that in these cases "the paper should be modified or not published."[12] This challenge remains unresolved as there are no agreed-upon guidelines for determining when a paper should be modified or when it should not be published.

Around the same time, a National Research Council (NRC) report helped frame the debate about open scientific communication in the life sciences. *Biotechnology Research in an Age of Terrorism*, which became known as the Fink Report after study committee chair Gerald Fink, highlighted a concept of dual use research through its identification of the "dual use dilemma in which the same technologies can be used legitimately for human betterment and misused for bioterrorism." The concept of applying the results of research undertaken for one purpose to other, sometimes controversial, ends was not new. But life scientists were much less familiar with addressing security concerns than their colleagues in the physical sciences and engineering.[13] The Fink Report argued for preparedness and made a series of recommendations on the oversight of research that raised potential security concerns. The recommendations drew on existing regulations, provided guidelines, and leveraged the traditions of self-governance in the life sciences. The report stressed the need to grapple with potential dual use risks early in the research process:

> By the time a manuscript is submitted for publication, substantial information about the research may have already been disseminated through informal professional contacts, presentations of preliminary results at scientific meetings, or consultations with colleagues. This is why the Committee recommends a system that can address research at its earliest stages, and why it is so important to

---

[12] Journal Editors and Authors Group, "Statement on Scientific Publication and Security," *Science Online*, February 21, 2003, Vol. 299, No. 5610, p. 1149.

[13] This is a result, in part, of the implementation of the Biological Weapons Convention and the ban on the development of biological weapons. The resulting cessation of acknowledged state biological weapons programs made these weapons appear less relevant to life sciences researchers.

make scientists aware of their personal responsibilities to consider the balance of risks and benefits in research they consider undertaking. Nevertheless, publication of research results provides the vehicle for the widest dissemination, including to those who would misuse them. It is thus appropriate to consider what sort of review procedures can be put in place at the stage of publication to provide another layer of protection.[14]

With regard to publication, the report endorsed self-governance by the scientific community.[15] It also endorsed the principles laid out in a national security decision document issued during the Reagan era [National Security Decision Directive 189 (NSDD-189)] which provided that, unless the work is classified, open release consistent with statutory requirements was the appropriate course.[16] The report also noted that, to be effective,

> any process to review publications for their potential national security risks would have to be acceptable to the wide variety of journals in the life sciences, both in the United States and internationally. . . . Continued discussion among those involved in publishing journals—and between editors and the national security community—will be essential to creating a system that is considered responsive to the risks but also credible with the research community.[17]

In 2005, in response to recommendations made in the Fink Report, the U.S. government established the National Science Advisory Board for Biosecurity (NSABB) to assist the federal government in assessing the potential risks of life sciences research and to offer advice to policymakers, research institutions, and researchers about the conduct, oversight, and communication of sensitive research.[18] As almost all research in the life sciences could potentially be considered "dual use" and to underscore that only a small set of experiments raise significant issues, the NSABB created a new category of research it described as "dual use research of concern" (DURC):

> Research that, based on current understanding, can be reasonably anticipated to provide knowledge, products, or technologies that could be directly misapplied by others to pose a threat to public health and safety, agricultural crops and other plants, animals, the environment, or materiel.[19]

---

[14] *Biotechnology Research in an Age of Terrorism*, pp. 116-117.

[15] "Publication of research results provides the vehicle for the widest dissemination, including to those who would misuse them. The Committee believes strongly that this part of the system should be based on the voluntary self-governance of the scientific community rather than formal regulation by government." See *Biotechology Research in the Age of Terrorism*," p. 8.

[16] NSDD-189 is discussed further herein.

[17] *Biotechnology Research in an Age of Terrorism*, p. 117.

[18] Information about the National Science Advisory Board for Biosecurity (NSABB) may be found at http://osp.od.nih.gov/office-biotechnology-activities/biosecurity/nsabb.

[19] *Biotechnology Research in an Age of Terrorism*, p. 17.

CONTINUING CONTROVERSIES OVER DISSEMINATION

Since its inception, the NSABB has been asked by a few federal agencies to review several manuscripts of concern (see Table 1-1). The manuscripts included a 2005 paper that described research conducted to reconstruct the influenza virus responsible for the 1918 Spanish Flu epidemic that claimed 40 to 50 million lives across the globe. When the manuscript was undergoing peer review for publication in *Science*, it was "recognized that the work might raise questions about the propriety of publication" and the authors were urged to consult experts at the Centers for Disease Control and Prevention, the U.S. National Institute of Allergy and Infectious Diseases, and the Office of Biotechnology Activities at the National Institutes of Health. Concerns were subsequently raised by the Office of the Secretary of the U.S. Department of Health and Human Services. This prompted a request for the members of the NSABB to review the paper.[20] The NSABB approved the paper for publication but suggested changes (see Table 1-1). The paper[21] was then published with an accompanying editorial in *Science*, but without the textual changes recommended by the NSABB.[22]

Another controversial 2005 paper provided a mathematical model of a potential bioterror attack on the food supply through the introduction of botulinum toxin into the milk supply.[23] The paper was approved for publication in the *Proceedings of the National Academy of Sciences* (PNAS) and the authors' uncorrected proof was provided under embargo to reporters, but publication was delayed, and the embargo extended, in response to a letter from the Assistant Secretary for Public Health Emergency Preparedness of the U.S. Department of Health and Human Services. PNAS and National Academy of Sciences representatives met with government representatives to discuss their specific concerns about the paper. Following this meeting, the Council

---

[20] See D. Kennedy, "Better Never Than Late," *Science*, October 14, 2005, Vol. 310, No. 5746, doi:10.1126/science.310.5746.195, p. 195.

[21] T. M. Tumpey et al., "Characterization of the Reconstructed 1918 Spanish Influenza Pandemic Virus," *Science,* October 7, 2005, Vol. 310, No. 5745, pp. 77-80.

[22] In a subsequent issue of *Science*, Editor-in-Chief Donald Kennedy wrote an editorial about the government's authority to restrict publication and the role of the NSABB specifically. "Government officials can advise," he wrote, "and should be listened to thoughtfully. But they can't order the nonpublication of a paper just because they consider the findings 'sensitive.' No such category short of classification exists, as the Reagan-era Executive Order National Security Decision Directive 189, still in force, makes clear. If a paper should not be published because of biosecurity risks, then it should be classified. Second, the NSABB should regard this first exercise as a helpful one-off and turn to its mandate of developing principles rather than making decisions on individual papers." See D. Kennedy, "Better Never Than Late," *Science*, October 14, 2005, Vol. 310, No. 5746, doi 10.1126/science.310.5746.195, p. 195.

[23] L. M. Wein and Y. Liu, "Analyzing a Bioterror Attack on the Food Supply: The Case of Botulinum Toxin in Milk," *Proceedings of the National Academy of Sciences of the United States of America*, July 12, 2005, Vol. 102, No. 28, pp. 9984-9989.

TABLE 1-1 Manuscripts Reviewed by the National Science Advisory Board for Biosecurity (NSABB)

Manuscript Received by the NSABB	Date Received by the NSABB
T. M. Tumpey et al., *Characterization of the Reconstructed 1918 Spanish Influenza Pandemic Virus*	September 2005
J. K. Taubenberger et al., *Characterization of the 1918 Influenza Virus Polymerase Genes*	
J. J. Esposito et al., *Genome Sequence Diversity and Clues to the Evolution of Variola Virus*	November 2005
G. Garufi et al., *Sortase-conjugation Generates a Capsule Vaccine That Protects Guinea Pigs against Bacillus anthracis*	November 2011

NSABB Conclusions/Recommendations	Outcome
<ul><li>The papers should be published.</li><li>The authors should add language to elaborate on the public health benefits of the research.</li><li>The U.S. government should examine the issue of biocontainment practices for 1918 viruses.</li><li>A communication plan, including an editorial to accompany the publications, should be developed.</li></ul>	Published in *Science* and *Nature* respectively with an accompanying editorial
<ul><li>Communicate with the addition of appropriate contextual information (e.g., biosafety measures, public health benefits, rationale for decision to communicate)</li></ul>	Published in *Science*
<ul><li>As written, the findings described in the manuscript may indeed meet the criterion for dual use research of concern. However, the NSABB noted significant scientific deficiencies with the methodology and with the interpretation of the results of the research, and concluded that if the scientific deficiencies were appropriately addressed, the manuscript would likely not raise significant dual use concerns.</li><li>The NSABB noted the potential for the manuscript as written to be sensationalized and raise public concerns.</li><li>The NSABB provided additional observations and suggestions for possible revisions to the manuscript that were intended to help mitigate the potential for misunderstanding and sensationalism.</li></ul>	Published in *Vaccine*

continued

TABLE 1-1 Continued

Manuscript Received by the NSABB	Date Received by the NSABB
M. Imai et al., *Experimental Adaptation of an Influenza H5 HA Confers Droplet Transmission to a Reassortant H5 HA/ H1N1Virus in Ferrets*	November 2011
S. Herfst et al., *Airborne Transmission of Influenza A/H5N1 Virus Between Ferrets*	

Courtesy of Elisa D. Harris, University of Maryland.
SOURCE: National Institutes of Health Office of Science Policy, July 1, 2016.

NSABB Conclusions/Recommendations	Outcome
After review of the originally-submitted manuscripts, the NSABB recommended that: - Neither manuscript should be published with complete data and experimental details. - The conclusions of the manuscripts should be published but without experimental details and mutation data that would enable replication of the experiments. - Text should be added describing: 1) the goals of the research; 2) the potential benefits to public health (including informing surveillance efforts, pandemic preparedness activities, and countermeasure development and stockpiling efforts); 3) the risk assessments performed prior to research initiation; 4) the ongoing biosafety oversight, containment, and occupational health measures; 5) biosecurity practices and adherence to select agent regulation; and 6) that addressing biosafety, biosecurity, and occupational health is part of the responsible conduct of all life sciences research. - The authors submit a special communication/commentary letter to the journals regarding the dual use research issue. After the review of revised manuscripts, the NSABB recommended: - The revised Imai manuscript should be communicated in full. - The data, methods, and conclusions presented in the revised Herfst manuscript should be communicated, but not as currently written. - The U.S. Government should continue to develop national (and participate in the development of international) policies for the oversight and communication of dual use research of concern. - The U.S. Government should expeditiously develop a mechanism to provide controlled access to sensitive scientific information.	After revision, published in *Nature* and *Science* respectively

of the National Academy of Sciences elected to publish the article as originally accepted with a commentary by Bruce Alberts, President of the National Academy of Sciences. Alberts suggested that the paper should "be used by the NSABB as a case study to help guide both the government and the scientific community in further matters of this kind."[24]

Subsequently, in 2011, the NSABB reviewed two papers submitted for publication in *Science* and *Nature*, respectively, by U.S. government-funded research teams in the United States and the Netherlands.[25] The papers identified genetic mutations that conferred aerosol-based mammalian transmissibility to H5N1 avian influenza, a highly pathogenic strain. The papers were particularly controversial due to broader concerns about pandemic influenza. They became the focus of international attention and put a spotlight on DURC research and the NSABB's role (see Chapter 2).

In all, the NSABB has reviewed six manuscripts of dual use concern between 2005 and 2012. While, to some, this suggests that there is not a significant problem, to others this suggests that problematic research is not being identified.[26] It is difficult to make an assessment either way as data on the number of papers rejected for publication (or modified prior to publication) on the basis of dual use concerns are not collected across journals.[27] Moreover, given the vital role that publishing plays in defining the success of a research career, there is a strong disincentive to impose restrictions at the time of publication. As such, leaving such decisions to the final stages of a research project is not ideal.

More recently, in 2013, researchers at the California Department of Health announced the discovery of a new strain of *Clostridium botulinum*. Botulinum

---

[24] B. Alberts, "Modeling Attacks on the Food Supply," *Proceedings of the National Academy of Sciences of the United States of America*, July 12, 2005, Vol. 102, No. 28, pp. 9737-9738.

[25] This particular case is particularly illustrative of the complicated nature of global research and publishing. One team was Japanese-American working in the United States with U.S. government funding. The other team was funded by the U.S. government but working in the Netherlands. The team based in the United States was seeking to publish in an American journal (*Science*). The team in the Netherlands was seeking to publish in a British journal (*Nature*).

[26] The reviews highlight the challenge in taking actions that might prevent the publication of beneficial research that contributes to the scientific literature or to public health and safety.

[27] See, e.g., D. Patrone, D. Resknik, and L. Chin, "Biosecurity and the Review and Publication of Dual-Use Research of Concern," *Biosecurity and Bioterrorism: Biodefense Strategy, Practice, and Science*, 2012, Vol. 10, No. 3.

The American Society for Microbiology (ASM), publisher of multiple journals that seek to "advance the microbiological sciences" (see https://www.asm.org/index.php/journals), reports that of the manuscripts submitted to the ASM journals *Antimicrobial Agents and Chemotherapy*; *Applied and Environmental Microbiology*; *Clinical and Vaccine Immunology*; *Infection and Immunity*; *Journal of Bacteriology*; *Journal of Clinical Microbiology*; *Journal of Virology*; *mBio*; and *Molecular and Cellular Biology*, those that mention DURC agents are 0.04% annually. Of the manuscripts submitted to these journals annually, "the total number of manuscripts rejected solely for DURC is 0%." Amy L. Kullas, Ph.D., Publishing Ethics Manager, Journals Department, American Society for Microbiology, communication with committee staff, March 15, 2017.

toxins are among those most dangerous to humans, and the researchers voluntarily opted not to release genetic information about the strain, as at that time, there was no known antidote for the newly discovered toxin.[28] Only later was it determined that the virulence of the strain could be blocked by available antitoxins.[29,30]

In view of ongoing concerns about the communication of biological research results that might convey significant risks and in the wake of incidents (such as the 2001 anthrax mailings)[31] in which biological materials were used for nefarious purposes, the United States has given significant attention to policies and practices that can enhance biosafety and biosecurity. A small but knowledgeable group of biological and social scientists, policy and security experts, and lawyers in the United States and overseas has become expert in various policy options to address biosecurity. However, most of the biological research community is not aware of these discussions and has not been actively engaged in them.[32]

CHARGE TO THE COMMITTEE

Our committee was charged with reviewing DURC policy and the management of DURC. Its objective was to review possible mechanisms for managing dissemination of research findings that strike an appropriate balance between the value of openness in scientific research and the needs of national security. As such, this encompasses the roles and responsibilities of students, researchers, institutions, publishers, and the federal government in the conduct of research. While one might think of dissemination in terms of publication, the committee, with encouragement from the project's sponsors, considered the management

---

[28] A commentary on the decision not to publish the information was included in the journal along with the article (see D. A. Relman, "'Inconvenient Truths' in the Pursuit of Scientific Knowledge and Public Health," *Journal of Infectious Diseases Advance Access*, October 7, 2013, Vol. 209, No. 2. Available at https://www.researchgate.net/publication/257535481_Inconvenient_Truths_in_the_Pursuit_of_Scientific_Knowledge_and_Public_Health.

[29] See H. Branswell "Researchers Keep Mum on the Botulinum Discovery," *Scientific American*, October 22, 2013. See also http://www.cidrap.umn.edu/news-perspective/2015/06/study-novel-botulinum-toxin-less-dangerous-thought.

[30] Unlike previous examples, this particular case is an example of basic research that generated new knowledge that raised concerns about dual use.

[31] Two sources of information about biosecurity incidents since 1900 are W. S. Carus, *Bioterrorism and Biocrimes: The Illicit Use of Biological Agents Since 1900* (Washington, DC: Center for Counterproliferation Research, National Defense University, 2001) and K. Berger et al., "Biosecurity Risk Assessment of Acts Targeting a Laboratory" in Gryphon Scientific, *Risk and Benefit Analysis of Gain of Function Research: Final Report—April 2016* (Takoma Park: Gryphon Scientific, 2016). Available at http://www.gryphonscientific.com/wp-content/uploads/2016/04/Risk-and-Benefit-Analysis-of-Gain-of-Function-Research-Final-Report.pdf.

[32] Individuals working with select agents and toxins or in particular fields, e.g., influenza research, would doubtless have knowledge of such discussions.

of dissemination as occurring along a spectrum from idea generation to the formal publication of research results in journals.[33] The committee hopes that the current report and its findings will provide policymakers with baseline information for further deliberation. Consequently, it does not provide recommendations for further action.

The committee gathered information both at a public information gathering meeting on July 11-12, 2016, and at a public workshop on January 4, 2017. To assist in its deliberations, the committee commissioned papers on a range of topics including biosafety and biosecurity, international approaches to biosecurity, ethics, export controls, and current government policies on information control. These papers are available at https://www.nap.edu/catalog/24761 under the Resources tab. Authors were asked explicitly to consider the implications of restrictions on the dissemination of scientific information, and it was expected that contrasting viewpoints would be expressed during the course of interaction between the committee, the audience, and session moderators. In Chapter 2, the committee provides a review of current U.S. policy and the broader international environment. Chapter 3 examines challenges and opportunities identified during the committee's meetings and in the commissioned papers. Chapter 4 offers findings to guide any reconsideration of DURC policy.

---

[33] Points along the spectrum include, for example, the point where research is funded, the period when research is being conducted, the transmission of information about research through informal communications among researchers, presentations at meetings and conferences, training and teaching, and the circulation of draft manuscripts and pre-prints or other self-published papers through traditional or electronic means.

2

The Current Policy Environment

In general, the U.S. government encourages and promotes the dissemination of basic research. The government also recognizes that some research may offer both benefits and risks and has, accordingly, developed policies to manage the dissemination of information in circumstances where it has authority. While acknowledging that its role has limits, the government recognizes that, given the nature of federal funding streams and the international scope of the life sciences research enterprise (see Box 2-1), there is significant value in frameworks and guiding principles that may be adopted by the larger community of researchers.[1]

The current U.S. government approach to the oversight of dual use research in general and dual use research of concern (DURC) in particular fits within the larger set of overlapping laws and regulations, policies, and guidelines that constitute the U.S. strategy for countering biological threats, including biological weapons and bioterrorism (see Figure 2-1).[2]

It is important to recognize that there are significant limitations to the reach of most regulations. In the particular case of DURC, the policies target research conducted with federal funding or at institutions that receive federal funding. They apply only to research that involves certain agents or pathogens and types of experiments. Moreover, the policies are aimed at seeking a level of oversight

---

[1] While their focus is research conducted in the United States, two National Academies' reports, Institute of Medicine and National Research Council, *Guidelines for Human Embryonic Stem Cell Research* (Washington, DC: The National Academies Press, 2005), doi:https://doi.org/10.17226/11278) and National Academies of Sciences, Engineering, and Medicine, *Human Genome Editing: Science, Ethics, and Governance* (Washington, DC: The National Academies Press, 2017), doi:https://doi.org/10.17226/24623, have provided guidelines that have wide applicability for the broader community of researchers. The guidelines have exerted particular influence internationally.

[2] National Security Council, *National Strategy to Counter Biological Threats* (Washington, DC, 2009). Available at https://obamawhitehouse.archives.gov/sites/default/files/National_Strategy_for_Countering_BioThreats.pdf.

BOX 2-1
Size of the Life Sciences Enterprise

While it is not possible to precisely calculate the size of the global life sciences enterprise, available data provide a good indication of the extent of activity in this area. In the United States, for instance, it is estimated that biotechnology generated $324 billion (2% of gross domestic product [GDP]) in revenue for the United States in 2012.[a] U.S. academic institutions spent $63.7 billion on research and development in all science and engineering fields in 2014. Of this total, the largest amount (59% or $37.6 billion) was spent on life sciences research.[b] 46% of publications in the United States are in medical and biological fields.[c] It is estimated that, in 2013, there were 21.1 million individuals in the United States with a bachelor's or higher level degree in a science or engineering field. Of this number, 2.4 million people held degrees in the life sciences.[d] The U.S. Bureau of Labor Statistics reports that, in 2014, there were 34,100 jobs for biochemists and biophysicists[e] and 79,300 jobs for biological technicians.[f]

Data suggest that the European bioeconomy generated about €2 trillion ($1.57 trillion) and employed more than 22 million people (approximately 9% of the European Union Workforce) in 2010.[g] In 2014, the EU's Gross Domestic Expenditure on R&D (GERD) as a percentage of GDP was 1.98%[h] of its GDP.[i] The EU has the largest percentage (22.2%) of the world's researchers,[j] with EU nations publishing more than 12,000 articles in the biological sciences in 2014.[k] The GERD spending for China, Korea, and Japan was 1.98%, 4.36%, and 3.35%, respectively.[l,m] These nations published more than 49,000 articles in the biological sciences in 2014.[n]

It is notable that a growing number of individuals unaffiliated with traditional research locations (e.g., academic institutions) are performing scientific experiments:

> "Do-It-Yourself Biology, or DIYbio, is a global movement spreading the use of biotechnology beyond traditional academic and industrial institutions and into the lay public. Practitioners include a broad mix of amateurs, enthusiasts, students, and trained scientists, some of whom focus their efforts on using the technology to create art, to explore genetics, or simply to tinker."[o]

While data on the movement are incomplete, according to a 2013 report, "the size of the DIYbio community is estimated at between 3,000 and 4,000 people, based on the DIYbio subscriber base and the estimates of community labs."[p] In response to a survey of the DIY community, 46% of respondents[q] indicated that they conduct work at a community lab, 35% indicated that they conduct work at hackerspaces,[r] 28% at academic, corporate, or government labs, 26% at home, and 8% at home exclusively.[s]

a R. Carlson, "Ubiquitous Biological Manufacturing." Available at http://sites.nationalacademies.org/cs/groups/pgasite/documents/webpage/pga_176175.pdf.
b National Science Board, *Science and Engineering Indicators*, 2016. Available at https://www.nsf.gov/statistics/2016/nsb20161/#/, p. 4. After the life sciences, the next greatest amount is spent on engineering (17% or $11 billion).
c Ibid, p. 5.
d Ibid.
e Bureau of Labor Statistics, "Occupational Outlook Handbook: Biochemists and Biophysicists." Available at https://www.bls.gov/ooh/life-physical-and-social-science/biochemists-and-biophysicists.htm. According to the Bureau of Labor Statistics, "Biochemists and biophysicists study the chemical and physical principles of living things and of biological processes, such as cell development, growth, heredity, and disease."
f Bureau of Labor Statistics, "Occupational Outlook Handbook: Biological Technicians." Available at https://www.bls.gov/ooh/life-physical-and-social-science/biological-technicians.htm. According to the Bureau of Labor Statistics, "Biological Technicians help biological and medical scientists conduct laboratory tests and experiments."
g Directorate-General for Research and Innovation, *Innovating for Sustainable Growth: A Bioeconomy for Europe* (Luxembourg: Publications Office of the European Union, 2012), p. 18. Available at http://ec.europa.eu/research/bioeconomy/pdf/bioeconomycommunicationstrategy_b5_brochure_web.pdf.
h See "OECD Science, Technology and Industry Outlook 2014." Available at http://www.oecd-ilibrary.org/science-and-technology/data/oecd-science-technology-and-r-d-statistics/oecd-science-technology-and-industry-outlook-2014_139a90c6-en.
i According to the World Bank, in current dollars, the 2014 GDP for the EU was $18.475 trillion (see http://data.worldbank.org/region/european-union); 1.98% of this is $366 billion.
j United Nations Educational, Scientific and Cultural Organization, *UNESCO Science Report: Towards 2030*, 2016, p. 33. Available at http://unesdoc.unesco.org/images/0023/002354/235406e.pdf.
k Ibid, p. 780.
l See "OECD Science, Technology and Industry Outlook 2014," OECD Science, Technology and R&D Statistics (database). Available at http://www.oecd-ilibrary.org/science-and-technology/data/oecd-science-technology-and-r-d-statistics/oecd-science-technology-and-industry-outlook-2014_139a90c6-en.
m According to the World Bank, in current dollars, the 2014 GDP for China was $10.482 trillion; the 2014 GDP for Japan was $4.849 trillion; and the GDP for Korea was $1.411 trillion (see http://data.worldbank.org/indicator/NY.GDP.MKTP.CD). R&D spending would thus have been $208 billion, $211 billion, and $47.3 billion, respectively.
n *UNESCO Science Report: Towards 2030*, p. 784.
o D. Grushkin, T. Kuiken, and P. Millett, *Seven Myths and Realities about Do-It-Yourself Biology* (Washington, DC: Woodrow Wilson International Center for Scholars, 2013), p. 4. Available at: http://www.synbioproject.org/site/assets/files/1292/7_myths_final-1.pdf.
p Ibid, p. 24.
q There were 305 respondents.
r Hackerspaces are community-operated workspaces for individuals with common interests.
s Survey respondents were given the opportunity to report that they conduct work at multiple locations. See *Seven Myths and Realities about Do-It-Yourself Biology*, p. 6.

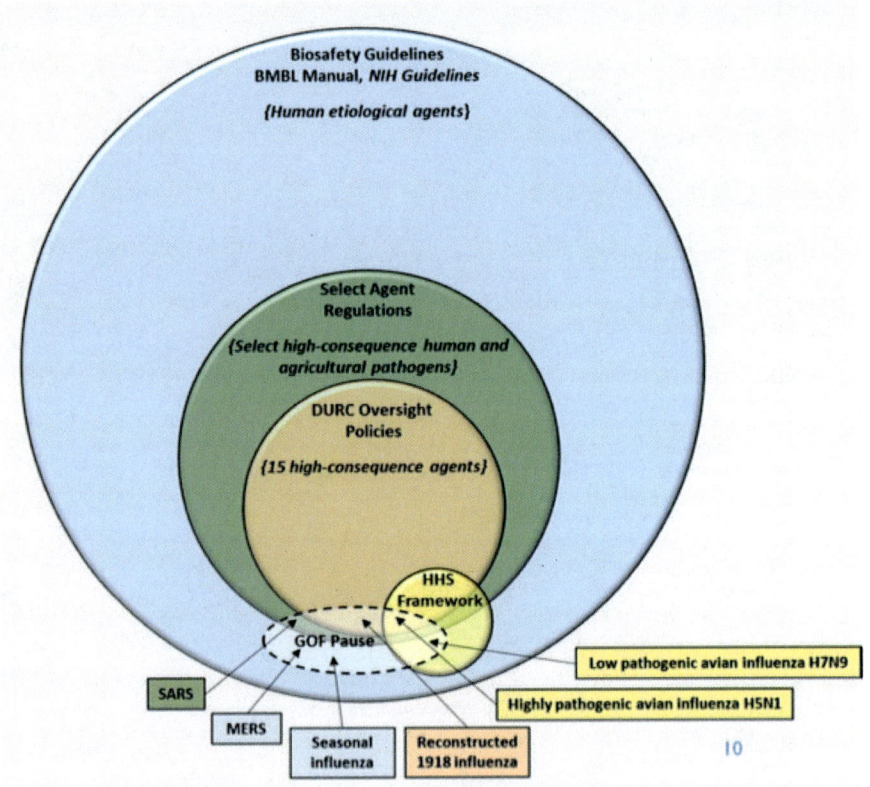

FIGURE 2-1 Comparison of the scope of different policies for the oversight of life sciences research involving pathogens.
NOTE: BMBL = Biosafety in Microbiological and Biomedical Laboratories; DURC = Dual Use Research of Concern; GOF = gain-of-function; HHS = U.S. Department of Health and Human Services; MERS = Middle East respiratory syndrome; SARS = Severe acute respiratory syndrome; NIH = National Institutes of Health.
SOURCE: National Science Advisory Board for Biosecurity, *Recommendations for the Evaluation and Oversight of Proposed Gain-of-Function Research* (Washington, DC, 2016), p. 28. This image is a work of the National Science Advisory Board for Biosecurity, taken or made during the course of an employee's official duties. As the work of the U.S. federal government, the image is in the public domain.

over research that could be used for dangerous purposes, not to prohibit such research or publication of research findings. These policies are complementary to other mechanisms that impose criminal penalties for misuse.

Policies governing the management of the dissemination of potentially sensitive scientific information focus on three categories of actors—research funders, research institutions, and researchers.[3] In addition, while not the subject of federal policies, scientific journals are critical players in the management of dual use research as they must make determinations about whether to publish potentially harmful research findings.

THE DEVELOPMENT OF U.S. GOVERNMENT OVERSIGHT OF INFORMATION WITH NATIONAL SECURITY IMPLICATIONS

By the early 1980s, concerns had grown that U.S. adversaries, in particular the Soviet Union, were using the openness of academic research in the United States to obtain militarily useful information. The National Research Council (NRC) undertook a study of security issues related to university research that resulted in the publication of *Scientific Communication and National Security* in 1982.[4] Known as the Corson Report, after study committee chair Dale Corson, the report concluded that for "the largest share [of university research], the benefits of total openness overshadow their possible near-term military benefits to the Soviet Union." The report noted that there are areas of research for which classification is clearly indicated and observed that there is a small "gray area" between openly disseminated research and classified research for which some controls might be appropriate.[5] The report does not explicitly consider the biological sciences.

In 1985, President Ronald Reagan issued National Security Decision Directive 189 (NSDD-189), which declared that "to the maximum extent possible, the products of fundamental research remain unrestricted . . . [and] where national security requires control, the mechanism for control of information

---

[3] See, e.g., U.S. Government, *United States Government Policy for Oversight of Life Sciences Dual Use Research of Concern*, 2012 (available at http://www.phe.gov/s3/dualuse/Documents/us-policy-durc-032812.pdf) and U.S. Government, *United States Government Policy for Institutional Oversight of Life Sciences Dual Use Research of Concern*, 2014 (available at http://www.phe.gov/s3/dualuse/Documents/durc-policy.pdf).

[4] Institute of Medicine, National Academy of Sciences, and National Academy of Engineering, *Scientific Communication and National Security* (Washington, DC: The National Academies Press, 1982), doi:https://doi.org/10.17226/253.

[5] Ibid, p. 4. The report (see p. 4) suggested that, for research in this area, "restrictions short of classification are appropriate." The report set forth criteria to be met before the communication of research could be restricted: the research was dual use or had direct military application; the technology was rapidly developing with a short time frame from the basic science to the development of an application; the dissemination of the research could give short-term military benefit to adversaries; and information about the technology was held only by the U.S. or friendly nations (see p. 5).

generated during federally funded fundamental research in science, technology and engineering at colleges, universities and laboratories is classification."[6] The Reagan Administration did not seek to introduce controls on "gray areas." NSDD-189 remains the foundation for U.S. policy related to potential restrictions on the publication of scientific research, and the directive has been reaffirmed by subsequent administrations, including by National Security Advisor Condoleeza Rice following the attacks of September 11, 2001, and most recently by the then Undersecretary of Defense Ashton Carter in 2010.

General U.S. government policy on classification is governed by federal statutes and executive orders. The most recent executive order (Executive Order 13526), issued in 2009, specifies that information may be classified if it is "owned by, produced by or for, or is under the control of the United States Government. . . . Basic scientific research information not clearly related to the national security shall not be classified."[7] There are only two limited exceptions. The government may classify even information that it does not own, does not control, or was not produced for it if: (1) the information is related to nuclear weapons (Atomic Energy Act of 1954), or (2) the information is a patent application whose disclosure "might . . . be detrimental to national security" (Invention Secrecy Act of 1951).[8] Controls on certain types of unclassified information are also laid out in specific statutes. These include sensitive security information,[9] unclassified controlled nuclear information,[10] and protected critical infrastructure information.[11]

---

[6] National Security Decision Directive 189 (NSDD-189): National Policy on the Transfer of Scientific, Technical and Engineering Information (September 21, 1985).

[7] White House Office of the Press Secretary, *Executive Order 13526—Classified National Security Information* (Washington, DC, 2009). Available at https://www.whitehouse.gov/the-press-office/executive-order-classified-national-security-information.

[8] The Invention Secrecy Act of 1951 (see https://www.govinfo.gov/content/pkg/STATUTE-66/pdf/STATUTE-66-Pg3.pdf) was enacted to prohibit the disclosure of inventions deemed by the Atomic Energy Commission, the Secretary of Defense, and the chief officers of defense agencies to be a detriment to national security. It allows for withholding of the granting of a patent if doing so is deemed to be in the national interest. Information on the number of patent secrecy order in effect at the end of fiscal year 2016 is available at https://fas.org/sgp/othergov/invention/.

[9] "Sensitive Security Information (SSI) is a control designation used by the Department of Homeland Security, and particularly the Transportation Security Administration. It is applied to information about security programs, vulnerability and threat assessments, screening processes, technical specifications of certain screening equipment and objects used to test screening equipment, and equipment used for communicating security information relating to air, land, or maritime transportation." See http://www.dhra.mil/perserec/osg/s2unclas/ssi.htm.

[10] Unclassified Controlled Nuclear Information (UCNI) is "certain unclassified information about nuclear facilities and nuclear weapons that must be controlled because its unauthorized release could have a significant adverse effect on the national security or public health and safety." See https://energy.gov/ehss/services/classification/unclassified-controlled-nuclear-information-ucni.

[11] Protected Critical Infrastructure Information (PCII) is "information not customarily in the public domain and related to the security of critical infrastructure or protected systems." See section 212(3) (6 U.S.C. 131(3)).

For information held by the government that is not classified or under specific statutory control, the Freedom of Information Act (FOIA) gives the public "the right to request access to records from any federal agency." "Federal agencies are required to disclose any information requested under the FOIA unless it falls under one of nine exemptions which protect interests such as personal privacy, national security, and law enforcement."[12]

Unless it is classified or subject to an exemption, data from research funded and held by the federal government are available under FOIA.[13] In fact, information in the possession of the federal government that is not classified or covered by a FOIA exemption is subject to release even if it is not owned by the government. Investigative journalists and others have used FOIA as a means to obtain information about institutional research, animal research, and biosafety records residing with agencies.[14] However, if the federal government is not in possession of federally funded information (e.g., DURC) held by researchers, it is not required to obtain such information from researchers in response to a FOIA request.[15]

Export controls are another mechanism by which dissemination of scientific information may be managed or restricted. As discussed in Chapter 3, these controls typically apply to the transfer of certain physical items and of non-public technical data associated with the items to particular destinations. Export controls may be brought to bear on information derived from research in two ways, but only if the transfer involves an export (the exchange of information among U.S. citizens in the United States is not limited or constrained by export controls). One involves the transfer of scientific information to foreign

---

[12] See https://www.foia.gov/about.html. The nine exemptions are: Exemption 1: Information that is classified to protect national security; Exemption 2: Information related solely to the internal personnel rules and practices of an agency; Exemption 3: Information that is prohibited from disclosure by another federal law; Exemption 4: Trade secrets or commercial or financial information that is confidential or privileged; Exemption 5: Privileged communications within or between agencies, including deliberative process privilege, attorney-work product privilege, or attorney-client privilege; Exemption 6: Information that, if disclosed, would invade another individual's personal privacy; Exemption 7: Information compiled for law enforcement purposes that: 7(A) could reasonably be expected to interfere with enforcement proceedings, 7(B) would deprive a person of a right to a fair trial or an impartial adjudication, 7(C) could reasonably be expected to constitute an unwarranted invasion of personal privacy, 7(D) could reasonably be expected to disclose the identity of a confidential source, 7(E) would disclose techniques and procedures for law enforcement investigations or prosecutions, or 7(F) could reasonably be expected to endanger the life or physical safety of any individual; Exemption 8: Information that concerns the supervision of financial institutions; Exemption 9: Geological information on wells.

[13] Carrie Wolinetz, National Institutes of Health, Presentation to the committee, July 11, 2016, New York, NY.

[14] See, e.g., A. Young and N. Penzenstadler, "Universities, Feds Fight to Keep Lab Failings Secret," USA Today, May 28, 2015. Available at https://www.usatoday.com/story/news/2015/05/28/labs-fight-for-secrecy/26530719/.

[15] State open access laws are also used to obtain information.

scientists, including even those working in the United States. The other relates to information redacted from a scientific publication. If a federal agency determines that aspects of a research study should not be exported, these aspects may be subject to export control regulations, such as requirements to obtain an export license before communicating the information to a foreigner.

In addition, the dissemination of scientific information can be controlled by the terms and conditions of a federal funder:

> Most government grants for unclassified technical activity specify that if the grantee believes the results of that work warrant classification, the grantee has the responsibility to limit the dissemination of that work and to contact the appropriate U.S. government agency that would have the authority to classify it. In such extraordinary cases, the initiative to seek classification rests with the grantee, not the government.[16]

THE EVOLUTION OF U.S. GOVERNMENT POLICY ON DUAL USE RESEARCH IN THE LIFE SCIENCES

After the September 11 attacks and the anthrax mailings that followed, the federal government enacted regulations to provide additional oversight for research on select agents and toxins.[17] The Uniting and Strengthening America by Providing Appropriate Tools Required to Intercept and Obstruct Terrorism Act of 2001 (PATRIOT Act) (P.L. 107-56) defined, in part, the reasons for which people may possess biological agents created through recombinant DNA technologies and specified those "restricted persons" who were not allowed to possess or transport select biological agents or toxins. In 2002, the Public Health Security and Bioterrorism Preparedness and Response Act (P.L. 107-188) added registration requirements for individuals working with biological select agents and toxins, added background checks for researchers, and provided additional guidance to the U.S. Department of Agriculture and the Department of Health and Human Services regarding listing select agents and safeguarding them during transfer. The Select Agent Program developed under the PATRIOT Act and the Public Health Security and Bioterrorism Preparedness and Response Act focuses on the handling of pathogens and toxins of concern; neither includes guidelines or requirements for the dissemination

---

[16] Commission on Scientific Communication and National Security, *Security Controls on Scientific Information and the Conduct of Scientific Research: A White Paper of the Commission on Scientific Communication and National Security* (Washington, DC: Center for Strategic and International Studies, June 2005), p. 6.

[17] Select agents and toxins are biological agents and toxins "which have the potential to pose a severe threat to public, animal or plant health or to animal or plant products." See https://www.selectagents.gov/. The current list of select agents and toxins is available at https://www.selectagents.gov/SelectAgentsandToxinsList.html.

of research results regarding the pathogens and toxins, although certain types of experiments require review before they are carried out.[18] The Select Agent Program requires reporting of "theft, loss, and release" of agents from laboratories registered with the program, and has made some of these data publicly available.

In 2005, based upon the recommendations of the Fink Report, the federal government established the National Science Advisory Board for Biosecurity (NSABB) to help assess the potential risks of life sciences research and offer advice to policymakers, research institutions, and researchers about the conduct, oversight, and communication of sensitive research.[19]

The NSABB was established to advise the Secretary of the Department of Health and Human Services, the Director of the National Institutes of Health (NIH), and heads of federal agencies that conduct, support, or have an interest in life sciences research. Originally, the board had a broad charge, but the current charge is much narrower (see Table 2-1). In particular, after the controversy over publication of work on influenza viruses with enhanced transmission properties, the April 2012 version of the charter removed the NSABB's capacity to "Review and provide guidance on specific experiments insofar as they exemplify a significant or particularly complex permutation of an existing category of dual-use research, or represent a novel category of dual-use research that requires additional guidance from the NSABB."[20]

NSABB activities during the board's first 10 years included issuing recommendations on the oversight of life sciences research of dual use concern in general[21] and on synthetic biology and gain-of-function (GOF) research involving pathogens with pandemic potential in particular. The NSABB also developed recommendations for codes of scientific conduct and encouraged a culture of

---

[18] Further information about the Select Agent Program may be found at http://www.selectagents.gov/. The original Select Agent Program was created in 1996 under the Antiterrorism and Effective Death Penalty Act (P.L.104-132) in response to the efforts of a researcher with ties to white supremacist organizations to obtain *Yersinia pestis* samples from the American Type Culture Collection. The program's purpose was to govern the transfer of an initial list of 42 pathogens and toxins. See J. E. Stern, "Larry Wayne Harris (1998)" in J. B. Tucker, ed., *Toxic Terror. Assessing Terrorist Use of Chemical and Biological Weapons* (Cambridge: MIT Press, 2000).

[19] Information about the National Science Advisory Board for Biosecurity may be found at http://osp.od.nih.gov/office-biotechnology-activities/biosecurity/nsabb.

[20] See G. D. Koblentz, "Is the NSABB Still Relevant to Today's Biosecurity Challenges?," July 16, 2014 (available at https://pandorareport.org/2014/07/16/is-the-nsabb-still-relevant-to-todays-biosecurity-challenges/) and Reuters, "US lawmakers question oversight of potentially dangerous experiments," August 13, 2014 (available at http://www.foxnews.com/health/2014/08/13/us-lawmakers-question-oversight-potentially-dangerous-experiments.amp.html).

[21] National Science Advisory Board for Biosecurity, *Proposed Framework for the Oversight of Dual Use Life Sciences Research* (Washington, DC, 2007).

TABLE 2-1 The Initial and Current Charge of the National Science Advisory Board for Biosecurity (NSABB)

2004 Charge	2016 Charge
• Develop criteria for identifying dual-use research and research results. • Develop guidelines for the oversight of dual-use research, including guidelines for the risk/benefit analysis of dual-use biological research and research results. • Provide recommendations on the development of a code of conduct for scientists and laboratory workers that can be adopted by professional organizations and institutions engaged in the performance of life science research. • Provide recommendations on the development of mandatory programs for education and training in biosecurity issues for all scientists and laboratory workers at federally-funded institutions. • Advise on national policies regarding the conduct of dual-use biological research. This includes strategies for addressing national security concerns while at the same time fostering continued rapid progress in public health research and food and agriculture research (e.g., new diagnostics, treatments, vaccines and other prophylactic measures, and detection methods). • Advise on national policies governing publication, public communication, and dissemination of dual-use research methodologies and results.	• Provide recommendations on the development of programs for outreach, education and training in dual use research issues for scientists, laboratory workers, students, and trainees in relevant disciplines. • Advise on policies governing publication, public communication, and dissemination of dual use research methodologies and results. • Recommend strategies for fostering international engagement on dual use biological research issues. • Advise on the development, utilization and promotion of codes of conduct to interdisciplinary life scientists, and relevant professional groups. • Advise on policies regarding the conduct, communication, and oversight of dual use research and research results, as requested. • Advise on the Federal Select Agent Program, as requested. • Address any other issues as directed by the Secretary of HHS.

TABLE 2-1 Continued

2004 Charge	2016 Charge
• Advise on national policies governing local review and approval processes for dual-use biological research, including the development of guidelines for the case-by-case review and approval by Institutional Biosafety Committees (IBCs).	
• Advise on criteria and processes for referral of classes of research or specific experiments by IBCs to the NSABB for guidance.	
• Review and provide guidance on specific experiments insofar as they exemplify a significant or particularly complex permutation of an existing category of dual-use research, or represent a novel category of dual-use research that requires additional guidance from the NSABB.	
• Respond to requests submitted by research institutions for the interpretation and application of the guidelines to specific research proposals in instances where a proposal has been denied by an IBC and the institution seeks additional advice.	
• Recommend strategies for coordinated international oversight of dual-use biological research.	
• Address any other issues as directed by the Secretary of HHS.[a]	

[a] See Congressional Research Service, *Oversight of Dual-Use Biological Research: The National Science Advisory Board for Biosecurity* (Washington, DC, April 27, 2007). "The NSABB is chartered for two-year intervals and [. . . continues] its work pending biennial renewals of the charter by the Secretary of the Department of Health and Human Services (HHS)." See National Institutes of Health Office of Science Policy, "NSABB FAQs." Available at: http://osp.od.nih.gov/office-biotechnology-activities/biosecurity/nsabb/faq.

responsibility, and educated scientists about DURC.[22] The board also engaged the international community around issues in life sciences DURC.[23]

As noted in Chapter 1, the NSABB refined the Fink Report's dual use concept in its 2007 *Proposed Framework for the Oversight of Dual Use Life Sciences Research: Strategies for Minimizing the Potential Misuse of Research Information*[24] to underscore that only a small set of life sciences experiments should raise significant issues. This resulted in the creation of the special category of research termed "dual use research of concern."[25]

The NSABB also addressed the communication and dissemination of research results, emphasizing the importance of monitoring research for dual use potential from experimental design through publication. In 2007, the board released a report titled *Responsible Communication of Life Sciences Research with Dual-Use Potential*.[26] That report offered an approach to "facilitate consistent decision making about the responsible communication of research information with dual use potential" and provided a "tool set" that includes: (1) principles for the responsible communication of research with dual use potential; (2) points to consider for identifying and assessing the risks and benefits of communicating research information with dual use potential; and (3) considerations for the development of a communication plan for research with dual use potential.[27] The NSABB also created two working groups on scientific journals' review policies and conducted surveys of journals' policies for reviewing DURC.[28]

The NSABB played a significant role in the 2011 H5N1 avian influenza (later described as the gain-of-function research) controversy and in the development of U.S. government policy for the oversight of DURC.[29] The NSABB provided an initial review of two controversial influenza papers in December

---

[22] C. Wolinetz, Presentation to the committee. See also "NSABB Reports and Recommendations" at http://osp.od.nih.gov/office-biotechnology-activities/biosecurity/nsabb/reports-and-recommendations.

[23] See http://osp.od.nih.gov/office-biotechnology-activities/biosecurity/nsabb/nsabb-meetings-and-conferences/international-engagement for further information.

[24] Available at http://osp.od.nih.gov/sites/default/files/resources/Framework%20for%20transmittal%20duplex%209-10-07.pdf.

[25] "Research that, based on current understanding, can be reasonably anticipated to provide knowledge, products, or technologies that could be directly misapplied by others to pose a threat to public health and safety, agricultural crops and other plants, animals, the environment, or materiel." See *Proposed Framework for the Oversight of Dual Use Life Sciences Research*, p. 17.

[26] National Science Advisory Board for Biosecurity, *Responsible Communication of Life Sciences Research with Dual Use Potential: A Set of Communication Tools Excerpted from the NSABB Proposed Framework for the Oversight of Dual Use Life Sciences Research*" (Washington, DC, 2007).

[27] Ibid, p. 3.

[28] C. Wolinetz, Presentation to the committee.

[29] See, e.g., G. Kwik Gronvall, *H5N1: A Case Study for Dual-Use Research* (New York: Council on Foreign Relations, July 2013). For a discussion of the GOF research controversy and the development of U.S. policy in this area, see Institute of Medicine and National Research Council, *Potential*

2011 and unanimously recommended against publication unless certain portions of the methods section were redacted. The board recommended that an arrangement be found to provide access to redacted information for researchers with a legitimate need.

The recommendations evoked a storm of controversy because they sought restrictions on publication.[30] Following additional discussion and receipt of additional information about the research and in light of the inability of the U.S. government to require that the redacted material be withheld,[31] the board voted in March 2012 to recommend publication of revised versions of both papers.[32] Concurrently, the U.S. government announced a policy to govern federal oversight of DURC.

The GOF controversy resulted in a number of policies for the oversight of DURC. In 2012, the U.S. government released the *United States Government Policy for Oversight of Life Sciences Dual Use Research of Concern*. The policy is intended to guide government agencies in the funding life sciences research. The policy applies to research involving one or more of 15 specific agents and toxins[33] and using one of seven types of experiments that: increase an agent or toxin's "harmful consequences;" disrupts immunity or effectiveness of immunizations to the agent or toxin; makes it resistant to prophylactic interventions or helps it evade detection; increases its stability, transmissibility,

---

Risks and Benefits of Gain-of-Function Research: Summary of a Workshop (Washington, DC: The National Academies Press., 2015), doi:https://doi.org/10.17226/21666.

[30] See, e.g., the discussions described in K. Matchett, A. Mazza, and S. Kendall, *Perspectives on Research with H5N1 Avian Influenza: Scientific Inquiry, Communication, Controversy: Summary of a Workshop* (Washington, DC: National Academies Press, 2013) and G. Kwik Gronvall, *H5N1: A Case Study for Dual-Use Research*.

[31] Francis Collins, at the time the director of the National Institutes of Health, informed the board that, upon the advice of counsel, the NSABB did not have the authority to redact manuscripts. See C. Wolinetz, Presentation to the committee; David A. Relman, Stanford University and VA Palo Alto Health Care System, Presentation to the committee, July 11, 2016, New York, NY; and Michael Imperiale, University of Michigan, Presentation to the committee, July 11, 2016, New York, NY.

[32] See National Science Advisory Board for Biosecurity, *National Science Advisory Board for Biosecurity Findings and Recommendations March 29-30, 2012* (Washington, DC, 2012) (available at http://osp.od.nih.gov/sites/default/files/resources/03302012_NSABB_Recommendations_1.pdf) and National Science Advisory Board for Biosecurity, *March 29-30, 2012 Meeting of the National Science Advisory Board for Biosecurity to Review Revised Manuscripts on Transmissibility of A/H5N1 Influenza Virus: Statement of the NSABB* (Washington, DC, 2012) (available at http://osp.od.nih.gov/sites/default/files/resources/NSABB_Statement_March_2012_Meeting.pdf). The board's vote in favor of publication was unanimous for one of the papers. For the other, the vote on publication was 12 in favor and 6 against.

[33] The agents and toxins are: 1) Avian influenza virus (highly pathogenic); 2) *Bacillus anthracis*; 3) Botulinum neurotoxin; 4) *Burkholderia mallei*; 5) *Burkholderia pseudomallei*; 6) Ebola virus; 7) Foot-and-mouth disease virus; 8) *Francisella tularensis*; 9) Marburg virus; 10) Reconstructed 1918 influenza virus; 11) Rinderpest virus; 12) Toxin-producing strains of *Clostridium botulinum*; 13) Variola major virus; 14) Variola minor virus; and 15) *Yersinia pestis*.

or ability to disseminate; alters its host range; makes a host population more susceptible to it; or "generates or reconstitutes an eradicated or extinct agent or toxin" on the list of 15 agents or toxins.[34] A researcher proposing to undertake experiments covered by the policy must make an initial assessment of potential risk and, if needed, develop, in collaboration with the federal funder, a risk mitigation plan.[35] The risk mitigation plan may limit the "venue and mode" of communication, or, if necessary, request voluntary redactions if risks cannot be adequately mitigated.[36] In extreme cases, the funding agency could elect not to provide funding, apply classification rules as a term and condition of funding, or terminate federal funding. According to the White House Office of Science and Technology Policy, "no Department or Agency has reported use of voluntary redaction of publication as part of a risk mitigation plan for any dual-use research of concern research project that has been reported to the Assistant to the President for Homeland Security and Counterterrorism pursuant to the" 2012 policy.[37]

The 2012 policy was supplemented by a 2014 policy that outlined the responsibilities of research-performing institutions receiving federal funding for life sciences research.[38] Researchers working with DURC were given a range of responsibilities including risk assessment and, in some cases, risk mitigation. Research institutions are required to fulfill their responsibility for the oversight of dual use research through an Institutional Review Entity that conducts reviews of institutional research of concern, develops risk mitigation plans, assesses and tracks compliance with the plans, and communicates information about activity of dual use concern to the funder of the research.

In addition, since February 2013, the Department of Health and Human Services (HHS) has conducted special reviews of requests for funding of GOF experiments involving highly pathogenic H5N1 avian influenza. The criteria used by HHS to evaluate proposed research are articulated in a 2013 HHS document titled *A Framework for Guiding U.S. Department of Health and Human Services Funding Decisions About Research Proposals with the Poten-*

---

[34] *United States Government Policy for Oversight of Life Sciences Dual Use Research of Concern.*

[35] For DURC that is proposed and not yet funded, departments and agencies will assess whether to incorporate risk mitigation measures in the grant, contract, or agreement. See *United States Government Policy for Oversight of Life Sciences Dual Use Research of Concern.*

[36] For a further discussion, see National Institutes of Health, *Tools for the Identification, Assessment, Management, and Responsible Communication of Dual Use Research of Concern: A Companion Guide to the United States Government Policies for Oversight of Life Sciences Dual Use Research of Concern* (Washington, DC: The National Academies Press, 2014), pp. 49-53. Available at https://www.phe.gov/s3/dualuse/Documents/durc-companion-guide.pdf.

[37] Gerald L. Epstein, Assistant Director for Biosecurity and Emerging Technologies, National Security and International Affairs Division, White House Office of Science and Technology Policy, communication with committee staff, June 19, 2017.

[38] *United States Government Policy for Institutional Oversight of Life Sciences Dual Use Research of Concern.*

tial for Generating Highly Pathogenic Avian Influenza H5N1 Viruses That Are Transmissible Among Mammals by Respiratory Droplets.[39] Two of the criteria are relevant to dissemination. One criterion is that "the research information is anticipated to be broadly shared in order to realize its potential benefits to global health," while the second is that the research "will be supported through funding mechanisms that facilitate appropriate oversight of the conduct and communication of the research."[40] The HHS document provides the only U.S. policy specific to GOF research.

Recent DURC policy and discussions have continued to focus on GOF research on pathogens with pandemic potential. Over time, the primary concerns about GOF research have shifted from biosecurity to biosafety. A series of serious biosafety lapses at federal laboratories during the summer of 2014, coupled with concerns about GOF research in light of the emergence of highly pathogenic avian influenza, severe acute respiratory syndrome (SARS) and Middle East respiratory syndrome (MERS), led the U.S. government to institute a pause in federal funding for certain GOF research.[41] The White House also instituted a deliberative process involving the NSABB. The board was tasked with developing recommendations for a system to provide oversight of GOF research. The National Academy of Sciences was asked to host two international symposia to collect broad stakeholder input for the NSABB's deliberations. The meetings explored the scientific and technical considerations involved in assessing the risks and benefits of GOF research and discussed the NSABB's draft recommendations.[42] The final NSABB *Recommendations for the Evaluation and Oversight of Proposed Gain-of-Function Research* were

---

[39] This document is available at http://www.phe.gov/s3/dualuse/Documents/funding-hpai-h5n1.pdf.

[40] See U.S. Department of Health and Human Services, *A Framework for Guiding U.S. Department of Health and Human Services Funding Decisions About Research Proposals with the Potential for Generating Highly Pathogenic Avian Influenza H5N1 Viruses That Are Transmissible Among Mammals by Respiratory Droplets* (Washington, DC, 2013). Available at http://www.phe.gov/s3/dualuse/Documents/funding-hpai-h5n1.pdf.

[41] White House, *U.S. Government Gain-of-Function Deliberative Process and Research Funding Pause on Selected Gain-of-Function Research Involving Influenza, MERS, and SARS Viruses* (Washington, DC, October 17, 2014). Available at http://www.phe.gov/s3/dualuse/Documents/gain-of-function.pdf.

The pause remains in place pending agency implementations of review mechanisms consistent with guidance issued in January 2017 (see *Recommended Policy Guidance for Departmental Development of Review Mechanisms for Potential Pandemic Pathogen Care and Oversight*, available at: https://www.phe.gov/s3/dualuse/Pages/GainOfFunction.aspx).

[42] Institute of Medicine and National Research Council, *Potential Risks and Benefits of Gain-of-Function Research: Summary of a Workshop* (Washington, DC: The National Academies Press, 2015), doi:https://doi.org/10.17226/21666; National Academies of Sciences, Engineering, and Medicine, *Gain-of-Function Research: Summary of the Second Symposium, March 10-11, 2016* (Washington, DC: The National Academies Press, 2016), doi:https://doi.org/10.17226/23484.

delivered to the White House in May 2016.⁴³ In January 2017, in response to the NSABB recommendations and an interagency review, the White House Office of Science and Technology Policy (OSTP) issued policy guidance recommending pre-funding review mechanisms for federal agencies that conduct or support the creation, transfer, or use of enhanced pathogens of pandemic potential (PPPs).⁴⁴ For federally sponsored research, the guidance recommends that agencies adopt the dissemination policies that currently apply to DURC research under the March 2012 DURC policy (i.e., "venue and mode" restrictions for risk mitigation, voluntary redactions, or classification).

A number of the NSABB's findings about U.S. policy for GOF research are relevant to the broader discussion of DURC policy. The 2016 report highlighted the overlapping policy and regulatory frameworks that provide oversight for DURC and certain types of GOF research. While only some of the policies and regulations are directly relevant to specific publication issues, most have the potential to impact dissemination. (See Box 2-2.)

THE INTERNATIONAL CONTEXT

Efforts to develop oversight mechanisms to manage the dissemination of DURC information in the United States take place in a broader international context.⁴⁵ In addition, the terms and conditions of U.S. government-funded research apply outside the United States, as the case of the NIH-funded researchers in the Netherlands embroiled in the 2011 GOF controversy illustrates. The United States may be global leader in biological research, however; absent a funding connection to the United States, U.S. DURC policies do not apply to activities in other countries.⁴⁶

---

⁴³ National Science Advisory Board for Biosecurity, *Recommendations for the Evaluation and Oversight of Proposed Gain-of-Function Research* (Washington, DC, 2016). Available at http://osp.od.nih.gov/sites/default/files/resources/NSABB_Final_Report_Recommendations_Evaluation_Oversight_Proposed_Gain_of_Function_Research.pdf.

⁴⁴ *Recommended Policy Guidance for Departmental Development of Review Mechanisms for Potential Pandemic Pathogen Care and Oversight.* Available at: https://www.phe.gov/s3/dualuse/Pages/GainOfFunction.aspx.

⁴⁵ This section of the report is intended to be descriptive. It does not speculate about how to manage research and publication that occurs outside the realm of U.S. influence or policy.

⁴⁶ See, for example, C. Rhodes, *International Governance of Biotechnology: Needs, Problems and Potential* (London: Bloomsbury Academic, 2010) and the discussions during the two Academies symposia on GOF research; Institute of Medicine and National Research Council, *Potential Risks and Benefits of Gain-of-Function Research: Summary of a Workshop* (Washington, DC: The National Academies Press, 2015), doi:https://doi.org/10.17226/21666 and National Academies of Sciences, Engineering, and Medicine, *Gain-of-Function Research: Summary of the Second Symposium, March 10-11, 2016* (Washington, DC: The National Academies Press, 2016), doi:https://doi.org/10.17226/23484.

> **BOX 2-2**
> **U.S. Government Policies Relevant to the Dissemination of Dual Use Research of Concern**
>
> NSDD-189 provides the foundation for current U.S. policy on the dissemination of scientific information. It provides that, "to the maximum extent possible, the products of fundamental research remain unrestricted . . . [and] where national security requires control, the mechanism for control of information generated during federally funded fundamental research in science, technology and engineering at colleges, universities and laboratories is classification."[a]
>
> In compliance with U.S. law and international legal obligations, the Departments of State and Commerce each administer a set of export control regulations that, in certain circumstances, impose restrictions on the flow of potentially dangerous biological information and materials.
>
> The Select Agent Program, administered by the Centers for Disease Control and Prevention and the U.S. Department of Agriculture, regulates federally and non-federally funded research involving the "possession, use, and transfer" of a prescribed list of human, plant, and animal pathogens.
>
> A U.S. government policy issued in March 2012 applies to research in the life sciences conducted or funded by federal agencies. A companion U.S. government policy issued in September 2014 covers the responsibilities of Principal Investigators and research-performing institutions, including the development and implementation of risk mitigation plans in consultation with the funding agencies.
>
> Since February 2013, the U.S. Department of Health and Human Services has conducted special funding reviews of gain-of-function experiments involving highly pathogenic H5N1 avian influenza.
>
> In January 2017, the White House Office of Science and Technology Policy issued policy guidance recommending pre-funding review mechanisms for federal agencies that conduct or support the creation, transfer, or use of enhanced pathogens of pandemic potential.
>
> ---
> [a] See National Security Decision Directive 189 (NSDD-189): National Policy on the Transfer of Scientific, Technical and Engineering Information (September 21, 1985).

As the Fink Report concluded in 2004:

> Although the focus of this report is on the United States, this country is only one of many pursuing biotechnology research at the highest level. The techniques, reagents, and information that could be used for offensive applications are readily available and accessible. And the expertise and know-how to use or misuse them is distributed across the globe. Without international consensus and consistent guidelines for overseeing research in advanced biotechnology, limitations on certain types of research in the United States would only impede the progress of biomedical research here and undermine our own national interests. It is entirely appropriate for the United States to develop a system

to provide oversight of research activities domestically, but the effort will ultimately afford little protection if it is not adopted internationally. This is a challenge for governments, international organizations, and the entire international scientific community. Efforts to meet that challenge are under way, but they must be quickly expanded, strengthened, and harmonized.[47]

There is no single international institution with the mandate or capacity to provide oversight of DURC, nor is any institution currently giving these issues systematic attention. International treaty organizations, United Nations Security Council (UNSC) resolutions, less formal international structures such as supplier agreements, and various components of international law could play a role in the management of DURC.

The core of the international biological nonproliferation and disarmament regime is the Biological and Toxin Weapons Convention (BWC), which was signed in 1972 and entered into force in 1975. It built upon the Geneva Protocol banning use of chemical and biological agents in war and was the first international disarmament treaty to ban an entire class of weapons.[48] The BWC prohibits development, production, stockpiling, and transfer of biological weapons, or the means of their delivery.[49] The BWC has provided a forum for discussions of dual use issues in the context of oversight of research (see Chapter 3), but it does not ban research on defensive measures.[50]

---

[47] National Research Council, *Biotechnology Research in an Age of Terrorism* (Washington, DC: The National Academies Press, 2004), p. 110.

[48] The formal title of the Geneva Protocol, which was signed in 1925 and entered into force in 1928, is the "Protocol for the Prohibition of the Use in War of Asphyxiating, Poisonous or Other Gases, and of Bacteriological Methods of Warfare." The Geneva Protocol prohibits first use of chemical and biological weapons. The BWC's formal title is the "Convention on the Prohibition of the Development, Production and Stockpiling of Bacteriological (Biological) and Toxin Weapons and on Their Destruction."

[49] The BWC states that "Each State Party to this Convention undertakes never in any circumstances to develop, produce, stockpile or otherwise acquire or retain:

(1) Microbial or other biological agents, or toxins whatever their origin or method of production, of types and in quantities that have no justification for prophylactic, protective or other peaceful purposes;
(2) Weapons, equipment or means of delivery designed to use such agents or toxins for hostile purposes or in armed conflict."

See U.S. Department of State, *Text of the Biological Weapons Convention*. Available at:https://www.state.gov/t/isn/bw/c48738.htm.

[50] States Parties to the BWC, of which there were 178 as of January 2017, are obligated to enact national implementing legislation in support of the treaty. Countries may have an array of laws and regulations that address biological weapons (as well as bioterrorism) directly or that contribute indirectly by governing various aspects of research and commercial activities. The U.S. implementing legislation for the BWC, the Biological Weapons Anti-Terrorism Act of 1989 (P.L. 101-298) is the primary means for law enforcement to take action and is not confined to a specific list of agents or toxins.

In 2004, to eliminate any potential gaps in the ability of the international regimes against weapons of mass destruction to respond to terrorism, the United Nations Security Council adopted UNSC Resolution 1540 (UNSCR 1540). The resolution obligates all United Nations (UN) members not to provide "any form of support to non-State actors that attempt to develop, acquire, manufacture, possess, transport, transfer or use nuclear, chemical or biological weapons."[51] UNSCR 1540 carries an obligation for UN member states to pass implementing legislation. Implementation is overseen by the standing 1540 Committee, but the committee has given limited attention to dual use research as a security issue. In December 2016, the UN Security Council unanimously adopted Resolution 2325. This resolution "encourages States, as appropriate, to control access to intangible transfers of technology and to information that could be used for weapons of mass destruction and their means of delivery."[52] This oversight is potentially relevant to the dissemination of DURC.

From time to time, other international organizations have become engaged in dual use issues, particularly the World Health Organization (WHO).[53] In 2005, for example, the WHO released a paper, *Life Science Research: Opportunities and Risks for Public Health*. This was followed by a workshop in 2006 on "Life Science Research and Global Health Security."[54] Additionally, a number of collaborative activities included regional workshops that addressed both biosafety and biosecurity issues. The final major WHO product prior to its involvement in the GOF controversy was a 2010 guidance document that provided a self-assessment tool for researchers and laboratories to evaluate their oversight of dual use research.[55] The recommendations of such documents are not binding on member states.

The WHO also became embroiled in the GOF controversy because of its role in global planning for influenza research. In February 2012, the WHO held

---

In remarks to the committee at its first meeting, Ed You, Federal Bureau of Investigation, stated that the bureau relies on the Biological Weapons Anti-Terrorism Act in its law enforcement activities.

[51] The text of the resolution may be found at http://www.un.org/en/ga/search/view_doc.asp?symbol=S/RES/1540(2004).

[52] See http://www.un.org/en/ga/search/view_doc.asp?symbol=S/RES/2325(2016).

[53] In May 1967, the WHO's World Health Assembly had approved a statement that "scientific achievements, and particularly in the field of biology and medicine—the most humane science—should be used only for mankind's benefit, but never to do it any harm." World Health Organization, World Health Assembly Resolution WHA20.54 (1967).

[54] World Health Organization, *Life Science Research: Opportunities and Risks for Public Health* (Geneva: World Health Organization, 2005) and World Health Organization, *Scientific Working Group on Life Science Research and Global Health Security: Report of the First Meeting* (Geneva: World Health Organization, 2007).

[55] World Health Organization, *Responsible Life Sciences Research for Global Health Security: A Guidance Document* (Geneva: World Health Organization, 2010).

a technical consultation of public health and influenza experts to review the manuscripts. The meeting concluded:

> On the question of limiting access to the results through publication of redacted versions, some participants observed that there was no current practical mechanism to limit access. Further, it would not be difficult for knowledgeable scientists to determine the information that had been removed, as novel methods had not been used. Limiting access to those with a need for the information would pose insurmountable practical problems. Chief among these problems are the development and implementation of a mechanism to disseminate the information to diverse and geographically distributed groups while maintaining the confidentiality of the detail. Therefore, such a mechanism would not realistically resolve concerns about dual-use research. There may be benefit in creating such a mechanism to deal with other dual-use research information in the future. However, this will require thorough consideration of and international agreement on practical issues such as security, access requirements, governance, and liability.[56]

The meeting was criticized by some who believed the meeting's outcome was predetermined by the large number of influenza virologists who were selected to participate. The issue of restricting access to some information from these manuscripts also threatened to undermine the Pandemic Influenza Preparedness (PIP) Framework created in 2011. PIP was designed to ensure improving sharing of influenza strains and information in light of concerns by countries such as Indonesia that they were not reaping the full benefits of their cooperation on global pandemic influenza preparedness efforts.[57]

As discussed further in Chapter 3, the WHO also held a larger conference in March 2013 to address broader issues of DURC policy, although the organization was largely consumed by the global Ebola crisis during the renewed U.S. policy debates that began in 2014. In principle, the WHO could, in the future, take up the issue of research oversight as it affects global health security. The WHO could provide an important complement to the BWC.

In addition to formal international treaties and international law, groups of states may organize themselves to undertake tasks or coordinate policy. In the biosecurity realm, the Australia Group (AG) is the most relevant. The AG was created in 1985 to improve consultation among member states on export controls. Originally focused on chemical weapons, the AG added biological weapons in the 1990s. The AG now has 41 member countries including the member states of the European Union and the EU as an institutional member. "[T]hrough the harmonisation of export controls, [the AG] seeks to ensure that exports do not contribute to the development of chemical or biological weapons." See http://australiagroup.net/en/.

---

[56] See http://www.who.int/influenza/human_animal_interface/mtg_report_h5n1.pdf?ua=1.
[57] See http://www.who.int/influenza/resources/pip_framework/en/.

The AG maintains common control lists for "dual use biological equipment and related technology and software, biological agents, and plant and animal pathogens" as the basis for promoting common standards and regulations. The second volume of the AG's *Common Control List Handbook* notes that controls on technology do not apply to information "in the public domain" (i.e., "technology that has been made available without restrictions upon its further dissemination"); "Controls also do not apply to 'basic scientific research,'" (i.e., "experimental or theoretical work undertaken principally to acquire new knowledge of the fundamental principles of phenomena or observable facts, not primarily directed towards a specific practical aim or objective") or "the minimum information necessary to apply for patents."[58] The handbook also describes several types of knowledge that may be controlled.

The AG has also provided a forum for discussion of issues, including dissemination of information. For example, when the Netherlands, an AG member, chose to rely on export controls as the mechanism for its oversight of the dissemination of the results of the Dutch research on H5N1 avian influenza, the AG held discussions on the GOF controversy.[59]

Many investments have been made by major donors to assist foreign countries with enhancements to both biosafety capacity and biosecurity (e.g., physical security, access controls, pathogen accounting, etc.).[60] Far fewer resources

---

[58] See Australia Group, *Common Control List Handbook, Volume II: Biological Weapons-Related Common Control Lists*, June 2014, p. 255. Available at http://www.defence.gov.au/export controls/_Master/docs/Australia_Group_Common_Control_List_Handbook_Volume_II.pdf.

[59] For an account of the Dutch experience with the GOF controversy, see K. van der Bruggen, "Biosecurity Challenges in the 21st Century: The Case of Gain-of-function Experiments," in S. Whitby, T. Novossiolova, G. Walther, and M. Dando, eds., *Preventing Biological Threats: What You Can Do* (Bradford: Bradford Disarmament Research Centre, 2015). Information about current Dutch policy, including export controls, is available from the Netherlands Biosecurity Office (see http://www.bureaubiosecurity.nl/en) and in the presentations in a side event during the 2015 BWC Meeting of Experts (the event may be found under "Side Events" at http://www.unog.ch/__80256ee600585943.nsf/(httpPages)/46cac219b57f8b49c1257db20030bce8?OpenDocument&ExpandSection=11#_Section11).

[60] Comprehensive data on international or U.S. expenditures on biosafety and biosecurity assistance are not available. One can gain a sense of the priorities from Defense Threat Reduction Agency (DTRA), *The Cooperative Biological Engagement Program Research Strategic Plan: Addressing Biological Threat Reduction Through Research,* 2015. Available at: http://www.dtra.mil/Portals/61/Documents/Missions/CBEP%20Research%20Strategy_FINAL_July%202015.pdf.

The largest U.S. assistance program, the Cooperative Biological Engagement Program's mission is to "establish and maintain international research collaborations with global partners to inform and enhance operational biosurveillance, enhance global health security, and foster safe, secure and sustainable bioscience capability with partner countries" (see p. 5).

The 5-year target for the Global Health Security Agenda's Biosafety and Biosecurity action package is broader: "A whole-of-government national biosafety and biosecurity system is in place, ensuring that especially dangerous pathogens are identified, held, secured and monitored in a minimal number of facilities according to best practices; biological risk management training and educational outreach are conducted to promote a shared culture of responsibility, reduce dual

have been devoted to awareness-raising, education and training, and policy development.

The result is that the call in the Fink Report for the establishment of an international effort to confront the challenge presented by DURC has not yet been met.[61]

CONCLUSION

Over the past 20 years, the U.S. government has developed, on the basis of the principles articulated in NSDD-189, a set of mechanisms, regulations, and policies to guide institutions and researchers conducting dual use life sciences research. These policies build on the fundamental idea that basic research should be open, but should be subject to policies that provide for the assessment and mitigation of risks in certain cases. Nevertheless, despite years of effort, there are some who still do not believe that the current federal approach is adequate to address concerns raised by current and emerging dual use research in the life sciences.

---

use risks, mitigate biological proliferation and deliberate use threats, and ensure safe transfer of biological agents; and country-specific biosafety and biosecurity legislation, laboratory licensing, and pathogen control measures are in place as appropriate." But its success is to be measured by the "number of countries who have completed/Completion of a national framework and comprehensive oversight system for pathogen biosafety and biosecurity, strain collections, containment laboratories and monitoring systems that includes identification and storage of national strain collections in a minimal number of facilities." See Global Health Security Agenda Action Packages at https://www.ghsagenda.org/packages/p3-biosafety-biosecurity.

There are programs that do address policy issues, including DURC, but they are significantly smaller. See, e.g., another DTRA program, The Project on Advanced Systems and Concepts for Countering WMD (https://www.usafa.edu/df/inss/indexpascc.cfm).

[61] Both the NSABB and the European Academies Scientific Advisory Council (EASAC) reports on GOF research, for example, call for greater efforts to engage the international community. The EASAC report explicitly references the Fink Report's recommendation of an International Forum on Biosecurity. See European Academies Scientific Advisory Council (EASAC), *Gain-of-Function: Experimental Applications Relating to Potentially Pandemic Pathogens. EASAC Policy Report 27,* (Halle: EASAC, 2015). Available at http://www.easac.eu/fileadmin/PDF_s/reports_statements/Gain_of_Function/EASAC_GOF_Web_complete_centred.pdf.

3

Managing Dual Use Research of Concern

Optimizing policies that encourage scientific openness and transparency while in appropriate cases limiting the dissemination of research results that might be applied to harmful ends is a difficult challenge.[1] There is significant debate among and within different communities about whether (and, if so, how) to limit the dissemination of such research results.[2] Numerous questions inform the debate. What types of information could be employed to cause harm? What criteria should be met before potentially harmful information is restricted? How widely should information be shared? What damage might particular information cause in the wrong hands? What materials and resources are necessary to translate information into a harmful application? Could the information be obtained through other means? What benefits might be foregone by not sharing information with those who could use it for legitimate purposes? What does the scientific community lose in quality control or follow-on work by withholding information from a wider audience?[3]

To assist it in answering questions such as these, the committee invited presentations and commissioned papers to explore options for the management of dual use research of concern (DURC). These papers are available at https://www.nap.edu/catalog/24761 under the Resources tab. The committee gathered information both at a public information gathering meeting on July 11-12, 2016, and at a public workshop on January 4, 2017. The July meeting consisted of presentations by invited experts, and the January workshop consisted

---

[1] For a discussion of this "wicked" problem, see, e.g., G. D. Koblentz, "Dual Use Research as a Wicked Problem, *Frontiers in Public Health*, August 2014, Vol. 2, Article 113, doi:10.3389/fpubh.2014.00113.

[2] Journal Editors and Authors Group, "Statement on Scientific Publication and Security," *Science Online*, 2003, Vol. 299, No. 5610, p. 1149; David A. Relman, Stanford University and VA Palo Alto Health Care System, Presentation to the committee, July 11, 2016, New York, NY; and Carrie Wolinetz, National Institutes of Health, Presentation to the committee, July 11, 2016, New York, NY.

[3] Gerald L. Epstein, White House Office of Science and Technology Policy, Presentation to the committee, July 11, 2016, New York, NY.

of presentations by the authors of the commissioned papers. At each event, members of the committee engaged presenters and members of the public in open discussions.

Presenters widely recognized the inherent tension between conducting life sciences research for the public good (e.g., to achieve economic, environmental, and public health benefits) and the need to increase awareness of risks associated with the fraction of life sciences research that could be directly misapplied to cause great harm. Acknowledging the limitations of current mechanisms for the management of life sciences research of concern, many expressed support for more effective policies and for guidance for researchers, research institutions, journal editors, and funders with regard to the conduct and dissemination of such research.[4]

FOUNDATIONAL U.S. POLICIES ON THE DISSEMINATION OF RESEARCH

National Security Decision Directive 189 (NSDD-189)

As noted in Chapter 2, NSDD-189 provides the foundation for current U.S. policy on the dissemination of scientific information. It states that "no restrictions may be placed upon the conduct or reporting of federally funded research that has not received national security classification, except as provided in applicable U.S. statutes."

In their paper commissioned for the committee, Michael Imperiale (University of Michigan) and David A. Relman (Stanford University and VA Palo Alto Healthcare Center) argued that much research currently falls into a gray area between fundamental research, which is to be shared openly, and sensitive research warranting classification. They noted that the Corson Report, whose recommendations were the basis for the creation of NSDD-189, identified four criteria for identifying types of research that fell into neither the fundamental research nor the sensitive research categories. Research in this category was identified as potentially needing some sort of restriction, including voluntary

---

[4] Other researchers believe that too many rules and regulations apply to (and ultimately hinder) the conduct of research. Many are documented in the 2016 National Academies' report, *Optimizing the Nation's Investment in Academic Research: A New Regulatory Framework for the 21st Century* [National Academies of Sciences, Engineering, and Medicine, *Optimizing the Nation's Investment in Academic Research: A New Regulatory Framework for the 21st Century* (Washington, DC: The National Academies Press, 2015), doi:https://doi.org/10.17226/21824]. The report notes, for example, that the lack of harmonization of select agent regulations across agencies decreases efficiency (see p. 185). It suggests that improvements to the current export control regime could "bring significant benefits to national security, to commerce, and to the economy, as well as to federally funded university research" (see p. 191).

withholding by the researcher.[5] They argued some life sciences research falls in the gray area and should potentially be subject to restrictions.[6]

Imperiale observed that the international threat environment has changed since NSDD-189 was adopted. At that time, the United States enjoyed a significant technological advantage over its adversaries. "Now that the research playing field has leveled out," he asked, how will the United States "stay one step ahead of those with nefarious tendencies?"[7] He suggested that "slowing down access to dangerous information temporarily" is an alternative to having such information "freely accessible immediately."[8]

Other workshop participants made the counter-argument that more emphasis should be placed on determining what is dangerous and on what can be addressed through either regulatory or technical means (versus keeping information secret and not taking counter actions). Further, they suggested that, now that the playing field has leveled out, other nations also are funding and publishing (or not publishing) information that has the potential to be misused.[9]

Export Controls

In certain circumstances, U.S. export control regulations can impose restrictions on the flow of potentially dangerous biological information and materials. Voluntary acceptance of restrictions on communications could make fundamental research results that might otherwise be freely disseminated subject to export controls.[10] The International Traffic in Arms Regulations (ITAR), administered by the U.S. Department of State Directorate of Defense Trade Controls, and the Export Administration Regulations (EAR), administered by the U.S. Department of Commerce, Bureau of Industry and Security, apply to the dissemination of life sciences research.[11] Of course, export controls are

---

[5] Michael Imperiale and David A. Relman, *Options for Management of Potentially Dangerous Information Generated by Life Science Research* (commissioned paper available at https://www.nap.edu/catalog/24761 under the Resources tab.)

[6] NSDD-189 did not address research in this "gray area." Imposing the type of restrictions described would require a revision to NSDD-189 or the adoption of a specific statute.

[7] Michael Imperiale, University of Michigan, Presentation to the committee, July 11, 2016, New York, NY.

[8] Imperiale and Relman, p. 10.

[9] In the particular cases of the mousepox and botulinum papers, some have suggested that scientific/technical information later came to light that demonstrated that both papers may not have been as useful for misuse as they first seemed. This knowledge would not have come to light had the information not been available to a diverse technical audience.

[10] Specifically, the subject arose during the controversy over the publication of the H5N1 avian influenza papers in 2011.

[11] Kimberly Strosnider, Doron Hindin, and Peter D. Trooboff, *The Role of Export Controls in Regulating Dual Use Research of Concern: Striking a Balance Between Freedom of Fundamental Re-*

inherently limited as a control mechanism because they can result in restrictions only for interactions that constitute exports. Exchanges of information among U.S. citizens within the United States do not constitute exports and are beyond the reach of the export control system.

In their presentation to the committee at its January 2017 workshop, Kimberly Strosnider and Doron Hindin (Covington & Burling LLP) discussed the ITAR's U.S. Munitions List and the EAR's Commerce Control List as they relate to controls on pathogens and toxins. Prior to December 31, 2016, Category XIV of the Munitions List controlled all "[b]iological agents and biologically derived substances specifically developed, configured, adapted, or modified for the purpose of increasing their capability to produce casualties in humans or livestock, degrade equipment or damage crops," as well as related technical data and defense services. However, effective December 31, 2016, this language has been replaced, clarifying that the ITAR control only the most highly sensitive pathogens that have been effectively weaponized through gain-of-function (GOF) intervention. Other pathogens are subject to U.S. export controls through the EAR.[12] The EAR's Commerce Control List includes "dozens of microbes, including all 15 DURC agents and those regulated by the Federal Select Agents Program, as well as certain related vaccines, immunotoxins, medical products, etc. and related technology."[13] Strosnider and Hindin described how export controls cover cross-border activity, such as transfers from the United States or transfers between foreign countries, as well as the release of controlled information within the United States or abroad to non-U.S. persons.

Both regulations complement NSDD-189 in that they have carve-outs for fundamental research. EAR exemptions include "published" information. They allow internet upload and prepublication review by co-authors, other researchers, and conference organizers. In their paper commissioned for the committee, Strosnider, Hindin, and Trooboff stated that, under the EAR, researchers can also "freely share their research with the public, such as by uploading their research results to the Internet, unabated by EAR controls."[14] This applies to research that is not otherwise restricted, such as, for example, through government security classification or by the terms of a federally funded research grant. The ITAR have a public domain exemption and do not cover unrestricted information released into the public domain via eight specific modes of release, including libraries, newsstands, open conferences in the United States, and others. A federal appeals court in September 2016 ruled that, unlike the EAR,

---

search and National Security (commissioned paper available at https://www.nap.edu/catalog/24761 under the Resources tab).

[12] Ibid, p. 5.

[13] Kimberly Strosnider and Doron Hindin, Covington & Burling LLP, Presentation to the committee, Washington, DC, January 4, 2017.

[14] Strosnider, Hindin, and Trooboff, p. 14.

the ITAR do not recognize an exemption for information released through the internet without government approval, where the information does not otherwise qualify for the public domain exemption.[15]

At the January workshop, Strosnider and Hindin argued that there is incongruity between the aims of export controls and the consequences of their being applied to the dissemination of research results. As an example, they explained that export controls do not apply to the importation of DURC to the United States or to release within the United States to U.S. persons within the United States, despite national security risks that can arise, and have in fact arisen, in such contexts. Moreover, export controls typically regulate items based on technical specifications or sensitivity of the item and, in some cases, based upon the purposes for which the item was developed. When export controls might apply to the results of unrestricted research, a key issue is whether the information has been or will be released to the public through a designated means. A researcher's decision to release such unrestricted and unclassified information to the public can remove EAR controls on publicly available material, regardless of the content of the communication. As a result, the sensitivity of the information does not control whether it is subject to export controls.

Thus, Strosnider and Hindin suggested that export controls are poorly suited to protect national security with respect to DURC that is imported into the United States or released within the United States to U.S. nationals, or with respect to unrestricted research results that are or will be properly released to the public.

Strosnider and Hindin further explained that broadening U.S. export controls to apply to privately generated research results could expose regulatory authorities to constitutional First Amendment claims. Such claims have been litigated in the past with respect to the ITAR, though the government has thus far prevailed—albeit by a narrow margin at times—with federal district

---

[15] In their paper (see p. 8), Strosnider, Hindin, and Trooboff address the particular case of publication of material on the internet and describe the 2016 ruling:

"a federal appeals court ruled that the Department of State, through the ITAR, could deny a U.S. citizen the ability to share privately-generated, unclassified information with the public through the Internet. The information at issue was a computer-aided-design file of a gun that would allow anyone to produce firearms using commercially available 3-D printers. In a 2-1 ruling, the court sided with the State Department, determining that the government's 'exceptionally strong interest in national defense and national security outweighs Plaintiff['s] very strong constitutional rights under these circumstances.' The court accepted the State Department's position that the ITAR's public domain exception was not available because the intangible and informal mode of dissemination of the computer file had failed to correspond to any of the ITAR's eight enumerated public domain provisions; had the data been published in materials available at a public library or newsstand, instead of the Internet, the ITAR public domain provision likely would have applied. The majority decision prompted a vigorous dissent," which "argued that the government's attempt to restrict uploading such information to the Internet 'appears to violate the [ITAR's] governing statute, represents an irrational interpretation of the [ITAR], and violates the First Amendment.'"

and appellate courts prioritizing national security interests over litigants' First Amendment rights.

Finally, Strosnider and Hindin suggested that asserting more stringent export control rules on research results could erode incentives for collaboration between the scientific research community and the U.S. government. As an example, they explained that increased enforcement or stricter export control rules might encourage researchers to avoid important prepublication national security reviews by relying on the EAR and ITAR public domain exceptions.

LIMITATIONS OF DURC POLICY

The 2012 and 2014 U.S. government policies for the oversight of life sciences DURC apply to U.S. government-funded research that involves the use of one of the 15 specified agents or toxins and one of the 7 categories of experiments. Several presenters noted that the result is policy that is simultaneously too broad and too narrow—not all research that involves the identified pathogens and experiments is research of concern and experiments outside of those on the listed experiments may raise dual use concerns. At the January workshop, Imperiale observed, "It is obvious that manuscripts involving pandemics are important, but what about new technologies—synthetic biology, gene drives, etc.? What about technologies that we haven't thought of yet?" As we move forward, he noted, there will always be new things to be concerned about.

It is necessary to weigh biosecurity risks against the benefits of free and open communication, which complicates determinations regarding the appropriate scope of restraints on dissemination. Such assessments currently are made on a case-by-case basis, by a variety of agencies and organizations. With no agreed upon common standard or process, the tradeoffs between biosecurity risks and the benefits of open communication continue to be debated. Under the current system, DURC policies could, in certain instances, place constraints on research that exceed the level of control necessary to serve legitimate biosecurity goals. On the other hand, current DURC policies might not constrain research that arguably should be subject to restriction.

Current DURC policies do not apply to classified research, research that does not involve 1 of the 15 specified agents and particular types of experiments, or to research at institutions that do not receive U.S. government funding.[16] Non-compliance presents the potential risk of the withdrawal of federal funding, but it is not clear whether other sanctions would, in fact, be imposed.

Much of the current policy is focused on formal publication. However, as noted in statement 4 of the journal editors' "Statement on Scientific Publication and Security" in 2003, scientific information is communicated by many other

---

[16] Elisa D. Harris, University of Maryland, Presentation to the committee, July 11, 2016, New York, NY.

means: "seminars, meetings, electronic posting, etc."[17] Traditionally, scientific results were published primarily in printed journals, and these journals were only readily accessible to subscribers or at institutions holding subscriptions to the journals. Even when journal articles began to be posted online, they were accessible only to those with subscriptions. An open access movement and the proliferation of public digital libraries has subsequently made much scientific literature widely available. Furthermore, researchers increasingly choose to post pre-prints of their research to internet servers before peer review.[18] Furthermore blogs, conferences, and widespread internet communications have meant that research results are often made available long before formal publication. Moreover, while "journals and scientific societies can play an important role in encouraging investigators to communicate results of research in ways that maximize public benefits and minimize risks of misuse,"[19] there is no mutually-agreed-upon approach to decisions surrounding the publication of DURC findings. The diverse channels through which information may be shared present challenges for the development of any policy attempting to manage dissemination.

At the committee's July 2016 meeting, Philip Campbell (*Nature*) reported that, in 2012, the editors of *Nature* decided that, as a general policy, the journal would not redact key findings or distribute information only to selected recipients. He suggested that redacting key data or methods disables subsequent research and peer review and that the distribution of redacted information to a select group of people on a need-to-know basis is practically infeasible because of questions such as: "Who holds the data?"; "Which criteria are used to determine who is allowed to see the redacted information?"; "Who decides by which mechanisms is the information then made accessible?"; and "How can information distributed to a university or public health laboratory remain confidential?" He suggested that biosecurity constraints on publication risk eroding the robustness of the field if reproducibility is not tested. Further, he suggested that delays and uncertainty about the ability to publish and to get credit may discourage young scientists from entering the field.

Campbell said that *Nature* has had a few papers of dual use concern since the 2011 GOF controversy. "There are six examples of such papers from 2015

---

[17] Journal Editors and Authors Group, "Statement on Scientific Publication and Security." See also A. S. Fauci and F. S. Collins, "Publishing Risky Research," *Nature*, May 2, 2012, Vol. 485, No. 5.

[18] For a discussion of the use of pre-prints in the biological sciences, see P. D. Schloss, *Preprinting Microbiology*, doi:https://doi.org/10.1101/110858. Available at http://www.biorxiv.org/content/early/2017/03/15/110858.article-info.

[19] Journal Editors and Authors Group, "Statement on Scientific Publication and Security."

and 2016" for which a technical assessment was seen as needed. In each case, he said, the outcome was "that no paper was rejected on the basis of risk."[20]

Inder Verma (*Proceedings of the National Academy of Sciences of the United States of America*) raised the issue of pre-print servers. What, he asked, should be done with material deposited into pre-print servers, where information is deposited and available publicly before it receives external review?[21]

Imperiale stated that he believes that neither reviewers nor editors are properly trained to identify DURC. He suggested that few individuals are qualified to make appropriate decisions about DURC and argued that it is unfair to request that journals screen for DURC manuscripts, as they do not have the proper experience.[22] He suggested that we "change the status quo and encourage [funding agency] responsibility by identifying potential DURC projects upfront and com[ing] up with a proactive plan."[23] In his paper commissioned for the committee, Tim Stearns (Stanford University) stated that he also believes

---

[20] Philip Campbell, *Nature*, Presentation to the committee, July 11, 2016, New York, NY.

A wider literature illustrates how few journals have policies to address dual use research. A 2009 survey of 28 major life sciences journals found that "few of the English-language publishers and none of the Russian and Chinese publishers surveyed implement formal biosecurity policies or inform their authors and reviewers about potentially sensitive issues in this area." See J. van Aken and I. Hunger, "Biosecurity Policies at International Life Science Journals," *Biosecurity and Bioterrorism*, April 2009, Vol. 7, No. 1. In 2011, an even wider survey found that, "of the 155 journals that responded" (a 39% response rate), "only 7.7% stated that they had a written dual-use policy and only 5.8% said they had experience reviewing dual-use research in the past 5 years." See D. B. Resnik, D. D. Barner, and G. E. Dinse, "Dual-Use Review Policies of Biomedical Research Journals," *Biosecurity and Bioterrorism*, March 2011, Vol. 9, No. 1. A 2012 survey "of 127 chief editors of life sciences journals in 27 countries to examine their attitudes toward and experience with the review and publication of dual-use research of concern" found that "very few editors (11) had experience with biosecurity review, and no editor…reported having ever refused a submission on biosecurity grounds. Most respondents (74.8%) agreed that editors have a responsibility to consider biosecurity risks during the review process, but little consensus existed among editors on how to handle specific issues in the review and publication of research with potential dual-use implications." See D. Patrone, D. Resnik, and L. Chin, "Biosecurity and the Review and Publication of Dual-Use Research of Concern," *Biosecurity and Bioterrorism*, September 2012, Vol. 10, No. 3.

[21] Inder Verma, *Proceedings of the National Academy of Sciences of the United States of America*, Presentation to the committee, July 11, 2016, New York, NY.

[22] M. Imperiale, Presentation to the committee, July 11, 2016, New York, NY.

[23] Ibid. Prior to the promulgation of the *United States Government Policy for Oversight of Life Sciences Dual Use Research of Concern* in 2012, a number of agencies (e.g., the National Institutes of Health, the Department of Homeland Security, the Department of Defense) had policies in place to review their intramural research for potential DURC. The 2012 policy required *all* "federal departments and agencies that conduct or fund life sciences research" to "conduct a review to identify all current or proposed, unclassified intramural or extramural, life sciences research projects" involving specific select agents and types of experiments to determine whether they met the definition of DURC. See *United States Government Policy for Oversight of Life Sciences Dual Use Research of Concern*.

that there "are relatively few practicing scientists with sufficient background to assess [. . . DURC], and to engage with the relevant government personnel."[24]

Were a federal agency to determine that specific research poses a risk, current DURC policy mandates that it must work with the researcher or institution to formulate a risk mitigation plan, which may include "determining the venue and mode of communication (addressing content, timing, and possibly the extent of distribution of the information) to communicate the research responsibly."[25] At the committee's July 2016 meeting, Harris noted that government agencies have different requirements for when the risk-benefit assessments must be carried out and risk mitigation plans developed.[26]

Carrie Wolinetz (National Institutes of Health) noted that the U.S. Department of Health and Human Services has developed a framework to guide funding decisions on proposals for research anticipated to generate Highly Pathogenic Avian Influenza (HPAI) H5N1 viruses.[27] The framework requires multi-disciplinary, department-level, pre-funding review and approval for research that is reasonably anticipated to generate certain avian influenza viruses that are transmissible in mammals via the respiratory route.[28]

EDUCATION AND TRAINING

U.S. guidance policies place substantial responsibility on Principal Investigators (PIs) and institutional review entities (IREs). Fulfilling the responsibilities requires significant awareness of the relevant issues, but there are serious questions as to whether these individuals or entities have opportunities to gain the expertise necessary to identify, assess, and mitigate communication risks. Furthermore, systematic mechanisms for sharing best practices and lessons learned do not exist.

In his paper, Stearns provided a description of the lack of awareness among his peers about DURC and security issues. Citing his own academic experience in a highly regarded biology department, he relayed how few faculty are familiar with the work of the National Science Advisory Board for Biosecurity (NSABB) or with DURC and security concerns in general: "it is possible to function at a very high level in the research community with essentially no engagement with this issue." Anecdotally, he noted that many of his colleagues believe that DURC "regulations are for researchers working on explicitly 'concerning'

---

[24] Tim Stearns, *Moving Beyond Dual Use Research of Concern Regulation to an Integrated Responsible Research Environment* (commissioned paper available at https://www.nap.edu/catalog/24761 under the Resources tab.)

[25] National Science Advisory Board for Biosecurity. *Responsible Communication of Life Sciences Research with Dual-Use Potential* (Washington, DC, 2007). See item 4.1.e.vii., p. 3.

[26] E. D. Harris, Presentation to the committee.

[27] See https://www.phe.gov/s3/dualuse/Pages/HHSh5n1Framework.aspx.

[28] C. Wolinetz, Presentation to the committee.

problems" and have the "tendency to view the government simultaneously as a welcome source of research funding, and an unwelcome source of burdensome regulations."[29] He also believes that many life scientists have little to no knowledge about the history of the development of nation-state biological weapons programs.

At the July 2016 meeting, Harris cited a 2008 paper [30] that indicated that few scientists had thought about dual use potential or had any experiences with biosecurity review.[31]

Stearns cited an inadequate articulation of the risk. "Neither 'adversary' nor 'malevolent purposes,'" he said, "is well-defined in most scenarios, and are taken to mean different things in different contexts, and by people with differing knowledge of the capabilities and intent of various potential adversaries."[32]

In their commissioned paper commissioned for the committee, Duane Lindner and Winalee Carter of Sandia National Laboratories described the laboratories' methodology for assessing possible risk associated with information generated by their research programs. They recognized the challenge posed by the rapid pace of change in science and biotechnology, "which can affect the risk/benefit calculus in sudden and discontinuous ways," and emphasized that "attention to establishing a culture that is aware of the risks and ready to help manage them is essential."[33]

Lindner and Carter acknowledged that, for researchers, the "natural enthusiasm about the benefit of specific work can lead to amplification, while lack of specific information about risk—information about actions by adversaries or careful and thoughtful assessment of potential negative consequences—can lead one to minimize or discount risk." They acknowledged that this dynamic may be less severe at institutions like theirs where most researchers have access to

---

[29] Stearns, p. 5.

[30] S. Whitby and M. Dando, "Effective Implementation of the BTWC: The Key Role of Awareness Raising and Education," Bradford Review Conference Paper No. 26, November 2010. Available at http://www.brad.ac.uk/acad/sbtwc/briefing/RCP_26.pdf.

[31] E. D. Harris, Presentation to the committee.

Similar findings were reported by the National Research Council [see National Research Council, *A Survey of Attitudes and Actions on Dual Use Research in the Life Sciences: A Collaborative Effort of the National Research Council and the American Association for the Advancement of Science* (Washington, DC: The National Academies Press, 2009), doi:https://doi.org/10.17226/12460] and in B. Rappert, ed., "Education and Ethics in the Life Sciences" (Australian National University E Press, 2010), available at http://press-files.anu.edu.au/downloads/press/p51221/pdf/book.pdf?referer=190).

[32] Stearns, p. 8.

[33] Duane Lindner and Winalee Carter, Sandia National Laboratories, *Control of Sensitive Information: Policy, Procedure, and Practice in a National Security Context* (commissioned paper available at https://www.nap.edu/catalog/24761 under the Resources tab).

fuller details about potential threats: "knowledge can help engender a culture of caution as appropriate."[34]

Several presenters suggested that a key way to increase the scientific community's awareness of DURC is to integrate national security issues into students and staff training, potentially through expanded programs covering biosafety and/or responsible conduct of research. Linder and Carter noted that "training can ensure that personnel can understand why information is sensitive, how to identify sensitive information, and what policies and procedures should be followed."[35] At the January 2017 workshop, Joseph Kanabrocki (The University of Chicago) urged training programs for researchers and staff.[36] In his paper commissioned for the committee, Sam Weiss Evans (Harvard Kennedy School of Government) recommended including training on biosecurity from the outset of students' careers, "incorporated within a broader curriculum on responsible research and innovation."[37]

Evans believes that the strongest change will come from efforts to promote, in the next generation of academic leaders in emerging technologies, the view that science and security are not mutually exclusive and then support efforts to achieve institutional change in the training of students.[38] He described three programs in synthetic biology that train students and early career professionals in responsible innovation in general or biosecurity in particular. The Synthetic Biology Leadership Excellence Accelerator Program (Synbio LEAP) brings a network of people from academia, industry, and government into broad discussions about responsible innovation and stewardship within synthetic biology. The Emerging Leaders in Biosecurity Initiative of the Johns Hopkins Bloomberg School of Public Health's Center for Health Security focuses more specifically on biosecurity. In the international Genetically Engineered Machines (iGEM) competition, where more than 6,000 students from 40 countries compete each year, the Human Practices committee has worked closely with the Federal Bureau of Investigation and other national organizations, industry, and academia to design a range of methods to both make students aware of security concerns in their work, and to structure the type of work they are allowed to do to avoid the most likely security-sensitive areas, such as a policy on the development of gene drives. Together these initiatives constitute models for ways to provide biosecurity training to students and researchers in the life sciences.

---

[34] Ibid, p. 7.
[35] Ibid, p. 7.
[36] Joseph Kanabrocki, The University of Chicago, Presentation to the committee, January 4, 2017, Washington, DC.
[37] Sam Weiss Evans, *The Construction of New Security Concerns in the Life Sciences* (commissioned paper available at https://www.nap.edu/catalog/24761 under the Resources tab).
[38] Ibid, p. 6.

Reflecting on the "maker" community of life sciences practitioners, Stearns stated in his paper that he believes that the number of those "who don't pass through the standard college or university training . . . [and] who are capable and came upon that capability independent of such training is relatively small." He suggested, however, that the number "is likely growing as part of the growth of the maker community." He suggested that "there are many opportunities to interact with this and related communities, and [that] there are some excellent examples of individual efforts in the government [that] can have a large effect."[39]

In his paper commissioned for the committee, Joseph Kanabrocki discussed how, at The University of Chicago, "all research staff involved with" select agent research "are committed to the ethical and responsible conduct of science."[40] He described their code of conduct, one element of a culture of awareness about biosafety, which is signed annually by life sciences researchers who work with select agents. Kanabrocki suggested that similar codes of conduct be expanded to embrace biosecurity concerns related to DURC. He argued that such codes of conduct would heighten researchers' awareness of DURC and would encourage them to explore alternative experimental approaches that would generate the desired information through less risky means.[41]

A general code of conduct used at Kanabrocki's institution includes items relating to scientific and personal integrity (intellectual honesty, transparency in conflicts of interest, fairness in peer review, etc.) and technical responsibilities around safety protocols and various training requirements. Unlike the code of conduct for those working with select agents, researchers are not required to sign the general code of conduct. Kanabrocki believes that their signature should be required.

One item of the code relates to research with dual-use potential: "responsibilities include protection of potentially sensitive information and awareness of reporting and publication requirements associated with research with dual use potential."[42] Kanabrocki noted that the code of conduct is taken seriously by researchers and is embedded in a research culture in which regularly scheduled meetings offer a place for questions and discussion of potential biosafety issues. An expansion of this point could integrate biosecurity concerns into a proven

---

[39] Stearns, p. 10. Stearns cited, as an example, the efforts of the Federal Bureau of Investigation to connect with the "homebrew" communities, though, for instance, its sponsorship of and involvement in the International Genetically Engineered Machine (iGEM) competition.

Stearns used the terms "maker community" and "homebrew community" as synonymous for the DIYbio community (see Box 2-1). Technically, these terms describe different DIY communities.

[40] Joseph Kanabrocki, *Biosafety, Biosecurity and Dual-Use Research of Concern* (commissioned paper available at https://www.nap.edu/catalog/24761 under the Resources tab).

[41] Ibid, p. 11.

[42] J. Kanabrocki, Presentation to the committee.

structure already in place for biosafety. Kanabrocki noted, however, that not all institutions have such systems in place.

With regard to laboratory practice, Kanabrocki believes that, "lab accidents and laboratory-acquired infections [are] under-reported and the opportunities for sharing our best practices are missed as a result." Further, "Lessons that are learned through the investigations of accidents and injuries or illnesses should be shared so that we can learn from each other's mistakes."

While data on laboratory safety are incomplete,[43] biosafety and biosecurity data are available for research involving select agents. Those data show that where laboratory workers are provided with rigorous training, the safety and security record is very good. From 2004 to 2010, there were approximately 10,000 select agent investigators and among them there were: (1) no reports of theft; (2) one lost shipment (out of 3,412); and (3) 11 laboratory-acquired infections, with no fatalities or secondary infections. Of the 11 infections, the majority occurred outside of high-containment facilities in laboratories that did not customarily work with highly pathogenic organisms (i.e., diagnostic and BSL-2 laboratories) and where workers may lack training in handling such organisms.[44]

Recent biosafety incidents have sparked efforts to improve biosafety training and practices. A series of biosafety lapses at U.S. government laboratories in 2014, for example, led to the creation of a trans-federal task force that issued an array of recommendations designed to improve biosafety practices and foster a strong culture of responsibility.[45] The implementation of those recommendations engaged multiple agencies in addressing the problems.

Biosafety and biosecurity experts nonetheless continue to express concern about the lack of consistent and systematic reporting of biosafety—and biosecurity—lapses.[46] The reporting of biosafety errors and accidents provides a basis for continual learning and improvement. A lack of national requirements for reporting and sharing data about such "near misses" represents a lost opportunity to promote the best possible biosafety practices. Indeed, the

---

[43] Standard references are the studies cited in Chapter 1, footnote 6.

[44] J. Kanabrocki, Presentation to the committee.

[45] See Federal Experts Security Advisory Panel, *Report of the Federal Experts Security Advisory Panel* (Washington, DC, December 2014) and White House, *Next Steps to Enhance Biosafety and Biosecurity in the United States* (Washington, DC, 2015). Both documents are available at https://www.phe.gov/s3/Pages/default.aspx.

[46] In addition to the discussion at the committee's January 4, 2017, workshop, see the accounts of the two international symposia on the GOF controversy organized by the Academies at the request of the White House: Institute of Medicine and National Research Council, *Potential Risks and Benefits of Gain-of-Function Research: Summary of a Workshop* (Washington, DC: The National Academies Press, 2015), doi:https://doi.org/10.17226/21666 and National Academies of Sciences, Engineering, and Medicine, *Gain-of-Function Research: Summary of the Second Symposium, March 10-11, 2016* (Washington, DC: The National Academies Press, 2016), doi:https://doi.org/10.17226/23484.

NSABB called for the creation of a national database to house such data.[47] As part of its review of the 2014 biosafety lapses in federal laboratories, the Federal Experts Security Advisory Panel (FESAP) recommended the creation of "a new voluntary, anonymous, non-punitive incident-reporting system for research laboratories that would ensure the protection of sensitive and private information, as necessary." At the time of this report, this was being developed and implemented in stages beginning with a pilot by the U.S. Department of Health and Human Services.[48]

Those working with select agents are required to report "theft, loss, and release" of agents from laboratories registered with the program. Some argue, however, that the punitive tone of select agent reporting requirements discourages individuals from sharing biosecurity "near misses" that would be valuable learning tools if they were more widely available.[49]

Lindner and Carter highlighted the importance of a culture of awareness about sensitive information at Sandia National Laboratories. While acknowledging that policies and procedures are important for ensuring effective risk assessment in a changing environment, they emphasized that these policies and procedures must be embedded in a culture of awareness about information management in order to be effective. The authors noted that "all institutions have policy, procedures, and cultures that control sensitive information of other types" and suggested that those structures be expanded to include information generated from DURC. Elements of their culture of awareness include signs throughout the laboratory spaces reminding workers of the presence of sensitive information, risks inherent in mishandling it, and researchers' individual responsibility; the regular dissemination of information about malicious attempt to gain access to sensitive information; and briefings to researchers about threats to sensitive information generated at the laboratories.[50]

Local, national, and international approaches can provide awareness-raising, education and training, and ongoing guidance and opportunities to share best practices and develop common approaches to manage the dissemination of

---

[47] See *Recommendations for the Evaluation and Oversight of Proposed Gain-of-Function Research*.

[48] *Report of the Federal Experts Security Advisory Panel*, p. 4, and *Implementation of Recommendations of the Federal Experts Security Advisory Panel (FESAP) and the Fast Track Action Committee on Select Agent Regulations (FTAC-SAR)*, October 2015, p. 8 (available at https://www.phe.gov/s3/Documents/fesap-ftac-ip.pdf).

[49] See, e.g., the presentation of Barbara Johnson in Institute of Medicine and National Research Council, *Potential Risks and Benefits of Gain-of-Function Research: Summary of a Workshop* (Washington, DC: The National Academies Press, 2015), doi:https://doi.org/10.17226/21666 and of Gavin Huntley-Fenner in National Academies of Sciences, Engineering, and Medicine, *Gain-of-Function Research: Summary of the Second Symposium, March 10-11, 2016* (Washington, DC: The National Academies Press, 2016), doi:https://doi.org/10.17226/23484.

[50] Lindner and Carter, p. 6, and Duane Lindner, Sandia National Laboratories, Presentation to the committee, January 4, 2017, Washington, DC.

scientific information.[51] Measures can include biosecurity modules in training courses for graduate students and others; frequent review of guidelines and the framework for oversight and regulation; and careful monitoring and reporting of situations in which misuse of biological materials might occur.[52]

Stearns suggested that it is necessary to "develop a better understanding of the effectiveness of the measures already taken to educate [researchers] about dual use issues" and of measures "that might be taken in the future." "There is very little data," he said, "about what scientists, broadly considered, and the public really understand in this domain, and how they think about some important issues."[53]

INTERNATIONAL OVERSIGHT OF DURC

In his paper commissioned for the committee, Piers D. Millett (Biosecure Ltd.) discussed international perspectives on DURC and potential channels for expanding discussions around DURC management. His overall view was that no international consensus exists on the need to address DURC. Indeed, he said, the subject has been largely ignored at the international level in recent years; expanding the discussion will require a concentrated effort. Millett's analysis was based on a review of work done as part of the Biological Weapons

---

[51] A general introduction and overview may be found in National Research Council, *Challenges and Opportunities for Education About Dual Use Issues in the Life Sciences* (Washington, DC: The National Academies Press, 2011), doi:https://doi.org/10.17226/12958. The National Academies have carried out an extensive program of international activities on dual use issues supported by the Department of State, including Education Institutes in Responsible Science in the Middle East/North Africa and South and Southeast Asia (see *Responsible Conduct in the Global Research Enterprise: An Educational Guide*, available at: http://www.interacademies.net/File.aspx?id=19789). An international version of the Academies' *On Being a Scientist* released in 2016 by the InterAcademy Partnership called *Doing Global Science* includes a discussion of dual use and biosecurity issues (see http://www.interacademycouncil.net/24026/29429.aspx). From 2013 to 2015, the European Union Chemical Biological Radiological and Nuclear Risk Mitigation Centres program ran a project to create networks of universities to raise awareness on dual use concerns (see http://www.cbrn-coe.eu/Projects/TabId/130/ArtMID/543/ArticleID/46/Project-18-International-Network-of-universities-and-institutes-for-raising-awareness-on-dual-use-concerns-in-bio-technology.aspx). For examples of activities in the United States, see a series of collaborative activities among the American Association for the Advancement of Science (AAAS) and the Federal Bureau of Investigation that can be found on the AAAS website at https://www.aaas.org/oisa/aaas-fbi.

[52] See, e.g., the findings and recommendations in several reports from the National Academies, including National Research Council, *Science and Security in a Post 9/11 World: A Report Based on Regional Discussions Between the Science and Security Communities* (Washington, DC: The National Academies Press, 2007), doi:https://doi.org/10.17226/12013; National Research Council, *Responsible Research with Biological Select Agents and Toxins* (Washington, DC: The National Academies Press, 2009), doi:https://doi.org/10.17226/12774; and National Research Council, *Challenges and Opportunities for Education about Dual Use Issues in the Life Sciences* (Washington, DC: The National Academies Press, 2011), doi:https://doi.org/10.17226/12958.

[53] Stearns, p. 10.

Convention (BWC) and by the World Health Organization (WHO), as well as a small, informal survey of national experts who have been involved in DURC discussions at an international level.[54]

Millett attributed low levels of engagement with DURC to: (1) limited awareness of the issue, (2) competing demands on countries' limited resources, (3) a sense that the issue is not relevant to most countries, and (4) suspicions in many developing countries that U.S. motivations for raising the issue of DURC are to protect its technological lead in the life sciences and deny access to a broad category of technologies and knowledge to other nations.

In light of this context, Millett described several starting points for discussion that he considered most productive for broadening international engagement with DURC. First, he urged that the issue not be portrayed as a zero-sum game in which every security benefit comes with a development cost. Rather, the relationship between development and security should be highlighted. Second, the role of biosecurity in safeguarding the bioeconomy, where there is increasing interest and investment globally, should be emphasized. Third, he encouraged the expansion of discussion of DURC to be a part of the entire spectrum of measures used to address biosecurity risks.

Speaking at the January 2017 workshop, Millett argued that, independent of policy decisions, technical discussions also need to take place. He specified that these need to be in good faith: "one thing that the United States could do to show leadership would be to express very early on a willingness to listen to the output of that discussion." He stated that the United States must recognize that its own policies may need to be revised as a result of further international discussions, "and saying very clearly at the start of that process a willingness to do that would . . . help to engender a real sense of buy-in and to demystify some of what the U.S. motivations might be."[55]

Millett described the past and present roles of the BWC and the WHO and considered their and other organizations' possible role as a future home for discussions of DURC. In the early 2000s, the BWC began to address research, in addition to its long-time focus on the development and acquisition of biological weapons. In the 2008 Meeting of States Parties to the BWC, parties were encouraged to "be alert to potential misuse of research, and assess their own

---

[54] Piers D. Millett, Biosecure Ltd., *Gaps in the International Governance of Dual-Use Research of Concern* (commissioned paper available at https://www.nap.edu/catalog/24761 under the Resources tab). The survey was circulated to experts from BWC delegations in 28 countries (Australia, Belgium, Canada, Finland, France, Georgia, Germany, Hungary, India, Indonesia, Ireland, Italy, Japan, Kenya, Liberia, Malaysia, Mali, Mexico, the Netherlands, Norway, Pakistan, Portugal, Russia, Sierra Leone, Spain, Switzerland, Ukraine, and the United Kingdom). In total, eight responses were received from Canada, Germany, the Netherlands, Pakistan, Portugal, Spain, Switzerland, and the United Kingdom.

[55] Piers D. Millett, Biosecure Ltd., Presentation to the committee, January 4, 2017, Washington, DC.

research for dual-use potential," "seek to stay informed of literature, guidance, and requirements related to dual-use research," and "provide concise, practical guidelines, including criteria to define sensitive research and identify areas of greatest risk."[56]

The BWC engaged more directly with DURC from 2012 to 2014, following the controversy about the two papers describing GOF mutations of H5N1 avian influenza viruses. The report from the 2012 meeting "express[ed] support for 'enhanced national oversight of dual use research of concern without hampering the fullest possible exchange of knowledge and technology for peaceful purposes'," and the 2013 report again articulated "the value of increased national oversight of DURC and highlighted the possibility of developing international approaches" to DURC management. This report also outlined a possible role for the BWC in facilitating the exchange of national experiences as a foundation for expanded international harmonization. The 2014 report summarized nations' common understandings of DURC and described key areas for future work; an appendix also included proposals for national measures for dealing with DURC, which had not found consensus.[57] Since that time, less attention has been paid to these issues.

Millett discussed the WHO's key role in the 2012 GOF controversy and the consultation on DURC that it hosted in 2013. This consultation identified key concerns surrounding DURC, gaps in existing management systems, and potential ways these gaps might be addressed. It concluded that DURC is an issue of relevance to all countries, affirmed the importance of oversight mechanisms, noted that oversight pertains to the entire research cycle, and considered that while the "establishment of a legally binding global agreement or regulation is theoretically possible, such an approach would be expensive, slow, likely impractical and would not necessarily yield the desired benefits," recommending instead "guiding principles, toolkits, best practices and other forms of technical assistance would help countries formulate their own policies and procedures for managing DURC." The WHO highlighted that "communication and continuing dialogue across a broad range of sectors and stakeholders are essential to create a culture of responsibility, cooperation and trust," including an exploration of different ways of assessing risk. Lastly, the WHO's consultation found that "awareness-raising, education and training on biosafety, biosecurity and DURC are essential not only for researchers but also for all sectors and stakeholders."[58] Millett noted that the 2013 findings in the WHO's consultation have not been followed up on, although the WHO has been active in discussions of DURC at BWC reviews.

---

[56] Quoted on Millett, p. 2.
[57] Millett, pp. 3-4.
[58] Quoted in Millett, p. 5.

A third organization mentioned by Millett as a possible home for international discussions around DURC was the United Nations Educational, Scientific and Cultural Organization (UNESCO), which, he noted, seems aware of dual use issues in the life sciences and carries out work on responsible research and innovation. Nonetheless, UNESCO has not assumed a leadership role in this area, and its interest in these issues has been inconsistent.

THE NATIONAL SCIENCE ADVISORY BOARD FOR BIOSECURITY

While the numerous reports of the NSABB offer recommendations to the broader community for the oversight of DURC, the board's role is to advise the government.[59]

Wolinetz remarked that, if an "investigator had no awareness of DURC policies . . . googled DURC" and emailed the National Institutes of Health Office of Science Policy and said, I have this manuscript "and am worried that it has DURC concerns" the manuscript could end up in "internal biosecurity committees and discussion groups and interagency groups that have dealt with DURC policies" and "potentially trigger an NSABB review." But she said this would "very much depend on the manuscript." If additional expertise were needed from other parts of the government, she said, that could be brought in, but the process is not transparent. Wolinetz said that this would "be an extremely rare situation" and that she does not "know that it makes sense to create a bureaucratic process for that situation."[60]

As noted in Chapter 1, since 2004, the NSABB has reviewed six manuscripts of dual use concern. The board's review of the two controversial H5N1 GOF manuscripts in 2011 led to the development of a framework for reviewing DURC that is based on risk-benefit analysis.[61] According to Wolinetz, the framework seeks to address several questions:

- Are there reasonably anticipated risks to public health and safety from direct misapplication of this information, i.e., is novel scientific information provided that could be intentionally misused to threaten public health or safety?
- Are there reasonably anticipated risks to public health and safety from direct misapplication of this information, i.e., does the information point out a vulnerability in public health and/or safety preparedness?

---

[59] At the time of the renewal of its charter following the 2012 GOF controversy, the NSABB was given a reduced advisory role.
[60] C. Wolinetz, Presentation to the committee.
[61] This is encapsulated in the *United States Government Policy for Oversight of Life Sciences Dual Use Research of Concern*.

- Is it reasonably anticipated that this information could be directly misused to pose a threat to agriculture, plants, animals, the environment, or materiel?
- If a risk has been identified, in what timeframe (e.g., immediate, near future, years from now) might this information be used to pose a threat to public health and/or safety, agriculture, plants, animals, the environment, or materiel?
- If the information were to be broadly communicated "as is," what is the potential for public misunderstanding, that is, what might be the implications of such misunderstandings (e.g., psychological, social, health/dietary decisions, economic, commercial, etc.)? For sensationalism? [62]

Imperiale and Relman argued that the criteria for triggering special consideration of research results need to be broader than those currently articulated by the NSABB in its May 2016 guidelines for GOF research, which are focused on pathogenic infectious agents, and should encompass as-yet-unknown situations in the future in other research areas, for example, synthetic biology and systems biology. They observed that the 2005 paper that modeled an introduction of botulinum toxin into the milk supply[63] provided a particularly "important case study because it did not involve wet lab research, but rather was a theoretical modeling study, and can be viewed as representative of an increasingly common type of research involving 'big data' and data mining tools." "Work of this type," they said, "typically arises outside of science research settings routinely subjected to biosafety and biosecurity oversight, and is typically undertaken by individuals unfamiliar with the history of biosafety guidelines."[64]

At the January 2017 workshop, Stearns agreed that the NSABB has failed to embrace all research that could potentially be of concern within its definition of DURC, including unpredictable developments in the life sciences, such as the genome editing tool CRISPR/Cas9 or research in gene therapy.

OPTIONS FOR THE FUTURE MANAGEMENT OF THE DISSEMINATION OF DURC

As mentioned in Chapter 1, the discussion of specific options for managing the dissemination of DURC takes place within the larger context of changing perceptions in the international scientific community about the appropriate

---

[62] C. Wolinetz, Presentation to the committee.
[63] L. M. Wein and Y Liu, "Analyzing a Bioterror Attack on the Food Supply: The Case of Botulinum Toxin in Milk," *Proceedings of the National Academy of Sciences of the United States of America*, July 12, 2005, Vol. 102, No. 28, pp. 9984-9989.
[64] Imperiale and Relman, p. 5.

balance between scientific freedom and the broader social responsibilities of science. Historically, freedom of inquiry has been an absolute value. It remains so for an important part of the scientific community.[65] However, given the complex ethical, legal, social—and security—issues posed by continuing scientific advances, there is increasing support for a view that scientific research must operate within a broader social context and that scientific freedom comes with important responsibilities. The struggle to develop effective policies for GOF research is illustrative of how these issues are playing out in the life sciences.[66]

Several presenters acknowledged the difficulty of arriving at clear criteria for what constitutes DURC, but identified elements that they believe are important. Imperiale and Relman, for example, suggested that criteria should be able to encompass all areas of the life sciences and that the "line will undoubtedly be context-dependent in many dimensions, including the area of the work, the availability of countermeasures against any potential dangers and the means to use them, and even the socio-political environment of the world at the time the work is performed." They acknowledged that "it is difficult to develop

---

[65] A striking example comes from the International Council for Science (ICSU), for decades one of the most staunch advocates for the primacy of scientific freedom.
"To address and promote both aspects [freedom and responsibility], ICSU established the Committee on Freedom and Responsibility in the conduct of Science (CFRS) in 2006. This Committee differs significantly from its predecessors that, since 1963, had focused on scientific freedom, in that it is explicitly charged with also emphasizing scientific responsibilities." [ICSU, *Freedom, Responsibility, and Universality of Science* (Paris: International Council for Science, 2014), p. 3. Available at http://www.icsu.org/publications/cfrs/freedom-responsibility-and-universality-of-science-booklet-2014/CFRS-brochure-2014.pdf.]

[66] Finding 5 of the NSABB's *Recommendations for the Evaluation and Oversight of Proposed Gain-of-Function* states:

Finding 5. There are life sciences research studies, including possibly some GOF research of concern [GOFROC], that should not be conducted because the potential risks associated with the study are not justified by the potential benefits. Decisions about whether specific GOFROC should be permitted will entail an assessment of the potential risks and anticipated benefits associated with the individual experiment in question. The scientific merit of a study is a central consideration during the review of proposed studies but other considerations, including legal, ethical, public health, and societal values are also important and need to be taken into account.

European Academies Scientific Advisory Council (EASAC), *Gain-of-Function: Experimental Applications Relating to Potentially Pandemic Pathogens* (*EASAC Policy Report 27*) (Halle: EASAC, 2015) offers the following findings:

3.2 Self-regulation and harmonisation
Self-regulation means that there are checks and balances on research agreed within the scientific community and does not mean that each researcher is free to decide for themselves what procedures to follow. (p. 17)

3.7 Publication of sensitive information
Scientific freedom is not absolute and the scientific community recognises that some information is sensitive. (p. 19)

clear criteria that broadly define a line that ought not be crossed" but identified an experiment—a deliberate attempt to isolate a mutant form of human immunodeficiency virus (HIV) that can be transmitted by aerosol route—which should *not* be performed. They stated that, as is the case in their hypothetical example, when risks are potentially high and the benefits nonexistent, "an experiment should not be performed solely because someone finds it intellectually interesting."[67]

Evans outlined two paradigms for understanding the relationship between the scientific enterprise and society. One holds that they are separate, where science produces objective knowledge that will lead to societal benefit and society intervenes only occasionally in order to regulate where there is a clear likelihood of research having harmful impacts. The other holds that science and society are mutually constitutive; science is not separate from society, and decisions about whether science benefits or harms society are often contested and irresolvable. He likened the two perspectives to a "Newtonian versus a Quantum view of biosecurity." In a Newtonian perspective, he said, "you have discrete, fully characterized entities that you can control in their movements based on a set of simple laws." With the Quantum perspective, "you have an entangled set of systems where measurement of the system changes the system itself, and therefore control is very often an indirect process." In a "Quantum" biosecurity environment, in other words, the *processes* by which we determine whether knowledge is a security concern heavily structure which concerns we are able to see, and we can never fully know whether, at a particular point in time, a particular piece of knowledge is a concern or not. Such determinations, Evans said, depend on the context, on who is using the knowledge, and how the knowledge interacts with other pieces of knowledge, resources, and intentions. Further, he said, indirect governance of a system like this means giving those who construct and use dual use knowledge (not just DURC) the tools to make their own determinations of how concerned we should be about potential security issues.[68]

Evans supported the creation of "networks for constructing security concerns" that would provide flexible governance for emerging DURC concerns and include the scientific community, government, nongovernmental organizations, and industry. He also noted that, among the recommendations issued from past reports from the National Academies of Sciences, Engineering, and Medicine on DURC, such as the Corson Report and Fink Report, the recommendations that were often not implemented were those that see science and

---

[67] Imperiale and Relman, pp. 5-6.
[68] Sam Weiss Evans, Harvard Kennedy School of Government, Presentation to the committee, January 4, 2017, Washington, DC.

society as mutually constitutive: "our institutional structures don't have the capacity to see the world in this way."[69]

In his commissioned paper, Evans described the eight communication principles put forth in the NSABB's *Responsible Communication of Life Sciences Research with Dual Use Potential* (see Table 3-1), as "emblematic of a way of constructing threats in biology that works well for DURC, but provide little help when considering dual use issues in the rest of the life sciences." He noted how principles one through three assume a "linear model of innovation" in which societal concerns are raised only at specific points in the process, if at all. He criticized the "combination of a linear model of innovation and a hard line between academic freedom and national security" because they lead to the viewpoint that the security concerns of life sciences research are "a zero-sum game between freedom and security" and can "be resolved by drawing a line in the innovation process where societal concerns like security can come in." He critiqued principle four's focus on the technical elements of risk assessment and its exclusion of political and broader public concerns.[70]

In place of the NSABB's eight principles, Evans suggested seven new ones—"principles for crafting new objects of security concern within the life sciences" (see Table 3-1)—noting that they share many elements with the 2006 NRC report *Globalization, Biosecurity, and the Future of the Life Sciences*.[71] These seven principles assert that "decisions about which [research] lines to pursue, as well as the actual conduct of research, are inextricably embedded in cultural, economic, political, and technical systems" and that "communities, not individuals, are best placed to determine the level of security concern around an area of research."

Imperiale and Relman observed that "there are events that may occur in the near or long term that could force a reactive response and a scheme for managing information that may not be productive."[72] They called for a "thoughtful, deliberate plan for managing information that will inevitably arise and pose major risks to humans, other animals, plants and their supporting ecosystems"[73] and enumerated some common arguments against proactive action and identified flaws in each. Some have made the argument that, since no act of terrorism has, to date, used biological materials, there is a lack of interest among malicious actors in biological attacks. This, Imperiale and Relman argue, is incor-

---

[69] S. W. Evans, Presentation to the committee.
[70] S. W. Evans, *The Construction of New Security Concerns in the Life Sciences*, p. 2.
[71] Institute of Medicine and National Research Council, *Globalization, Biosecurity, and the Future of the Life Sciences* (Washington, DC: The National Academies Press), doi: https://doi.org/10.17226/11567.
[72] These include accidental or deliberate release of an agent from a laboratory; a bioterrorist attack; an unexpected zoonosis by a highly virulent pathogen; or development of an additional transformative bioengineering technology. See Imperiale and Relman, p. 7.
[73] Imperiale and Relman, p. 7.

TABLE 3-1 Principles for the Dissemination of Life Sciences Research of Concern

NSABB Principles for the Responsible Communication of Research with Dual Use Potential	Evans' Principles for Crafting New Objects of Security Concern within the Life Sciences
1. The open and unfettered sharing of information and technologies has been a hallmark of the life sciences and has fostered a steady stream of scientific advances that underpin public health and safety, a strong and safe food supply, a healthy environment, and a vigorous economy.	1. Life science research has been used to improve public health and safety, as well as provide for a safe food supply, a healthy environment, and a vigorous economy, but it has also aided in the accidental and purposeful debilitation, death, and destruction of people, the economy, and the environment. Which of these are likely outcomes is not always clear at any stage of research.
2. Progress in the life sciences relies heavily on the communication of research findings so that the findings can be both validated and used for further research.	2. In rare cases where there is broad acceptance of the security concerns around an area of research, such as 'experiments of concern' done on Select Agents with federal funds, established procedures for conduct and oversight of research should be followed.
3. Life sciences research should be communicated to the fullest extent possible to ensure the continued advancement of human, animal, plant, and environmental health. Consequently, any restriction of scientific communication should be the rare exception rather than the rule.	3. For all lines of research taken, there are many that are not pursued. Decisions about which lines to pursue, as well as the actual conduct of research, are inextricably embedded in cultural, economic, political, and technical systems.
4. There is a need for reasonable balance in decisions about the communication of research with dual use potential. It is important to recognize the potential for the deliberate and malevolent misuse of dual use research findings and to consider whether the disclosure of certain information might reasonably pose a threat to national security (i.e., public health and safety, agricultural crops and other plants, animals, the environment, or materiel). If the communication of dual use research does pose potential security risks, the logical next step is a risk-benefit analysis of communicating the information.	4. Communities, not individuals, are best placed to determine the level of security concern around an area of research. A relationship of mutual trust and shared expertise should be fostered in particular between the life science and intelligence communities.

continued

TABLE 3-1 Continued

NSABB Principles for the Responsible Communication of Research with Dual Use Potential	Evans' Principles for Crafting New Objects of Security Concern within the Life Sciences
5. After weighing the risks and benefits of communicating dual use research findings, the decision regarding communication is not necessarily a binary (yes/no) one. Rather, a range of options for communication should be identified and considered. The options available will depend on the research setting (e.g., academia, government, or private). They could range from full and immediate communication, to delayed and/or modified communication, to restricted/no communication, and could be recommended singly or in appropriate combinations on a case-by-case basis, depending on the nature of the dual use finding and the potential risks associated with its communication.	5. Broader public debates about security concerns in research are not 'crises of trust in science'. Instead, they are opportunities to assess "societal preferences for principles of achieving consent to a technology, distributing liabilities, and investing trust in institutions."[a]
6. Paradigms for the responsible communication of research with dual use potential should also take into consideration that the communication of dual use research can occur at multiple points throughout the research process, that is, at points well upstream of the publication stage. Thus, it is important to apply principles and practices of responsible communication at these early stages as well.	6. We cannot expect researchers to be engaged in this conversation unless their training and aspirations include it from the beginning, and it is incorporated within a broader curriculum on responsible research and innovation. This must be clearly championed and internalized by their mentors and advisors.[b]
7. It is important to consider not only what is communicated but also the way in which it is communicated. Investigators and sponsors of research with dual use potential should recognize that the communication of certain dual use information is likely to raise biosecurity concerns, not only within the scientific community but also within the general public. Consideration should be given to the potential for public concern and misunderstanding and for sensationalism. Thought should be given to the need for the inclusion of contextual and explanatory information that might minimize such concerns and misunderstanding.	

TABLE 3-1 Continued

NSABB Principles for the Responsible Communication of Research with Dual Use Potential	Evans' Principles for Crafting New Objects of Security Concern within the Life Sciences
8. Public trust is essential to the vitality of the life sciences research enterprise. It has always been important for life scientists to participate in activities that enhance public understanding of their research. However, because of the potential for public misunderstanding of and concerns about dual use research, it is especially important that life scientists conducting research with dual use potential engage in outreach on a regular basis to increase awareness of the importance of the research and to reassure the public that the research is being conducted and communicated responsibly.	

[a] S. Rayner and R. Cantor, "How Fair Is Safe Enough? The Cultural Approach to Societal Technology Choice." *Risk Analysis,* 1987, Vol. 7, No. 1, pp. 3–9.

[b] M. J. Palmer, F. Fukuyama, and D. A. Relman, "A More Systematic Approach to Biological Risk." *Science*, December 18, 2015, Vol. 350, Issue 6267, pp. 1471–1473.

SOURCES: National Science Advisory Board for Biosecurity, *Responsible Communication of Life Sciences Research with Dual Use Potential* (Washington, DC, 2007) and Sam Weiss Evans, *The Construction of New Security Concerns in the Life Sciences* (commissioned paper available at https://www.nap.edu/catalog/24761 the Resources tab).

rect. Biological materials have indeed been used for nefarious purposes, e.g., in an attempt by the Rajneeshee to alter the outcome of an election in Oregon by contaminating salad bars with Salmonella bacteria in 1984 and in the mailing of live anthrax spores in the aftermath of the September 11 attacks. They also addressed the argument that since only a few experiments will generate dangerous information they can be dealt with individually as they arise. They observed that "the consequences of just one episode of deliberate misuse of information could be enormous."[74] Imperiale and Relman countered the argument that full control of information is currently impossible. While this, they said, may be true, it is possible to create policies to discourage people with malicious intent or slow their progress. Finally, they responded to the argument that sensitive information already in the published literature has not been misused: this, they suggested, "is akin to someone in 2000 stating that since no one has ever deliberately flown a commercial jet into a large building, we don't have to

---

[74] Imperiale and Relman, p. 11.

worry about it." In light of this, they supported "a more proactive strategy for addressing what is already a clear and pressing set of challenges."[75]

Several presenters favored a broad approach, in two respects. They believed that the process of establishing guidelines for DURC management should be inclusive, and that an important attribute of whatever process is ultimately chosen should be based on broad, diverse input in determining what information is sensitive and defining the level of risk.

Evans recommended that a "relationship of mutual trust and shared expertise should be fostered in particular between the life science and intelligence communities." He recommended that the NSABB resume its efforts to build a network between the intelligence community and journal editors. He noted that the relationship between law enforcement and the scientific community has not always been optimal but highlighted the Federal Bureau of Investigation's Weapons of Mass Destruction Directorate, which has become a resource for the scientific community's security concerns, and urged that it be strengthened, institutionalized, and studied for how it might be shared more broadly. He also urged that the Department of Commerce Bureau of Industry and Security's Emerging Technology Research Advisory Committee strengthen its focus on the life sciences, noting its current lack of expertise.[76]

Both Stearns and Relman noted that, in the past 20 years, efforts have been made to stimulate discussion between scientists and national security experts.[77] Relman noted, however, that the nation has not undertaken bridge-building between the two communities in a strategic, coordinated, or thoughtful way.[78]

Imperiale and Relman examined several possible mechanisms for controlling dissemination of research results. One was controlled unclassified information (CUI), a category implemented by George W. Bush in 2008 in response to a proliferation of types of sensitive but unclassified information. They found it inadequate, citing a lack of clarity around who would authorize and then manage the CUI, but believed that some features of the concept could be useful.

Lindner and Carter described their procedures for protecting sensitive information at a research enterprise in which the generation of sensitive research results is commonplace. Their point of departure was the notion that all people, in their daily lives, have access to sensitive information of various sorts and carry out frequent risk/benefit analyses to decide when and where to make it public. In professional situations, common sense is often supplemented by policies and training: "laws, policies, and procedures create a framework for management of sensitive information. Training and situational awareness—especially aware-

---

[75] Imperiale and Relman, p. 11.
[76] S. W. Evans, Presentation to the committee.
[77] D. A. Relman, Stanford University School of Medicine and VA Palo Alto Health Care System, Presentation to the committee, January 4, 2017, Washington, DC, and T. Stearns, Stanford University, Presentation to the committee, January 4, 2017, Washington, DC.
[78] D. A. Relman, Presentation to the committee.

ness of risk—help create an environment that establishes norms and practices for assessing sensitivity of specific information and for managing it."[79] They discussed laws and policies that guide their actions concerning classified and controlled unclassified information, situations without strong parallels in the broader scientific community, and described their "explicit attempt to create and re-enforce a culture in which our staff are equipped to make appropriate decisions as they handle and manage sensitive information," which does have parallels.[80]

Self-regulation is valued by the scientific community. In self-regulation, Imperiale and Relman suggested, each researcher would realize what information is "of unusual risk" and would "somehow either decide not to disseminate or self-censor in some way." "Scientists," they said, "always do this. We deliberately choose . . . we put certain things in papers and put things on blogs or not." However, self-regulation in the realm of DURC is not widely practiced, meaning that "a well-meaning person runs into a lot of problems when the system is not set up to deal with this sort of circumstance."[81]

Relman described the properties of a new system, whose "purpose is to simply guide research towards mitigating the risks that have now been revealed by this information so that it no longer is so risky, so that this inevitably failing effort to fully prevent dissemination can then be released and information made available to everyone deliberately." A new system "would enhance the dissemination of the information to those that were so designated as in need of access, could make good use of access, and . . . slow access to everyone else." It would apply to publicly and privately funded research and would be transparent, deliberative, standardized, international, and adaptive. It would rely on the expertise of people in the sciences, public health, security, policy, and ethics.[82]

Imperiale and Relman suggested the formation of a diverse group of people to handle the management of DURC when research generating potentially sensitive results emerges. They asked, "if information needs to be controlled, who controls it?" as risk mitigation measures are created and deployed. Relman and Imperiale highlighted the importance of participation and buy-in from critical constituencies, including "respected members of the scientific, policy, and security communities, as well as other representatives of the general public," who would solicit the input of scientific experts as needed. The group would be agile and responsive yet forward-looking: "Ideally, this group would appreciate the need in some cases for taking action far in advance of the generation of the information." The authors' recommendation is that one or more entities take responsibility for controlling the sensitive information. They suggested

---

[79] Lindner and Carter, p. 1.
[80] Lindner and Carter, p. 2.
[81] D. A. Relman, Presentation to the committee.
[82] Ibid.

that scientific societies (and the InterAcademy Partnership in particular) could fulfill this role.[83]

CONCLUSION

Several options for managing the dissemination of DURC emerged in the papers commissioned by the committee and through the course of its discussions and consideration of relevant external materials. Of all the issues related to the oversight of DURC, questions about limiting the dissemination of research are the most controversial, as many in the scientific and policy communities believe that any restrictions placed on scientific pursuits could harm the research enterprise by limiting knowledge that might have value in efforts to respond to significant public health crises. The committee recognized that the presented options would be contentious.

During the committee's discussions and review of materials, the following elements were raised as important in the effective management of DURC:

- Ongoing, interactive education and training of individuals in the broader life sciences community;
- Engagement with advisory bodies with monitoring and/or enforcement capabilities;
- International harmonization of policies and approaches;
- Engagement with extant or newly convened international entities;
- Uniform roles and responsibilities for publishers;
- Legislative, regulatory, or policy mechanisms positioned at critical stages of the dissemination process; and
- Increased engagement with the public.

Implementation may necessitate additional resources, the establishment of best practices, refinement of policies and guidances, adoption of new laws, broader stakeholder engagement, and appropriately positioned and empowered advisory bodies. A clearer understanding of risk and benefit and the tradeoffs associated with these options is necessary before policy can be successfully implemented. To aid this process, in the following chapter, the committee provides specific findings that may serve to advance discussions to develop approaches for the future management of DURC.

---

[83] Imperiale and Relman, pp. 9-10.
Founded in 1993 and expanded and re-launched in 2016, the InterAcademy Partnership (IAP) is a global partnership of more than 130 merit-based national and regional academies of science, engineering, and health, which aims to maximize the contributions of science toward understanding and solving the world's most challenging problems. Through this structure, IAP and its members are active in countries that constitute 95 percent of the world's population. For more information, see http://www.interacademies.org/.

4

Committee Findings

The committee considered expert presentations given before it, the content of commissioned papers and related external materials, and public discussions and engaged in private deliberations. It offers the following findings on the state of managing dissemination of dual use research of concern (DURC). It hopes that these findings provide a baseline for the development of principles that will, in turn, lay the framework for government policy for managing the dissemination of information about the conduct and results of DURC research by federal agencies, the research community, and the international scientific community. In alignment with its charge, the committee is not offering recommendations.

CONTEXT: CHANGES IN RESEARCH AND COMMUNICATION TOOLS

A confluence of factors—including advancing technologies and technical capabilities, globalization, rapid sharing of information, the changing nature of scientific publication, and the capacity and intent of some to cause harm—has led to concerns about the dissemination of scientific information that could be directly exploited for nefarious purposes.

Scientific information is disseminated through a wide range of means including education, training, presentations and posters at conferences, preprint servers, informal communications, patents, and formal publication. The prevalence of digital information and online transmission and storage of information related to dual use research also makes information increasingly vulnerable to hacking. Much of current policy, however, tends to focus on formal publication.

There are some oversight mechanisms in place to make decisions about the publication of information that might pose risks to biosecurity. To date, the number of instances where detailed review has occurred and the frequency with which information has been restricted (by voluntary redaction, use of export controls, etc.) is small.

Findings

1. In general, the United States has a solid record with regard to the safe conduct of biological research. Given the lack of a comprehensive reporting system, knowledge of the nature and full extent of biosafety and biosecurity incidents is incomplete. Nevertheless, the number of documented, publicly known incidents of serious biosafety errors or lapses of biosecurity at laboratories has been small.[1]
2. In the wake of concerns that biological materials could be used for nefarious purposes and the significant risks that communication of the results of some biological research might convey, the United States has given significant attention to policies and practices that can enhance biosecurity.
3. Even with regard to research that could be directly applied to bioterrorism, there are concerns about excessive restrictions on the free flow of information. Open dissemination of research findings, a fundamental principle of research practice, can serve to alert relevant communities to a risk, provide the foundation for the development of countermeasures, and establish the foundation for scientific advances that could have significant public health benefits.

U.S. GOVERNMENT POLICY

Many policies potentially apply to the dissemination of DURC. U.S. DURC policies provide structures for managing the dissemination of information about certain pathogens and types of experiments that raise biosecurity concerns, but they apply only to research that is conducted at institutions receiving federal funding. Non-compliance presents the potential risk of the withdrawal of federal funding, but it is not clear whether other sanctions would, in fact, be imposed.

Findings

4. The dissemination of life sciences information that may raise biosafety and biosecurity concerns is governed by fragmented policies and regulations.
5. Federal policies on DURC reach only a portion of the individuals conducting life sciences research. Those conducting research at institutions that do not receive federal funds (whether in private industry, in the "Do-It-Yourself" community, in other nations, etc.) are not bound

---

[1] The committee is not suggesting that errors and lapses are inconsequential, as it recognizes that a single lapse could have significant policy and public health consequences.

by these policies, but other regulations such as export control laws, could apply.
6. Research that might be considered as DURC can, in principle, be identified before it is carried out or during the course of work when an unusual finding is encountered. Policies for identification of DURC in early phases, with consequent actions (a decision not to fund the research, withdrawal of funding, classification, mitigation plans, etc.), are in place for some types of research. Intervention at an early stage is more appropriate and likely to be more effective than at the time of publication.
7. The current policy focus and definition of DURC do not capture biosecurity concerns in all relevant areas of life sciences research, especially those that are emerging (e.g., synthetic and systems biology, computational modeling, genome editing, gene drives, neuroscience, the isolation of new micro-organisms and toxins). On the other hand, the current system of DURC policies and regulations may constrain certain types of research [e.g., research with select agents and toxins, research with pathogens of pandemic potential (PPPs)] more than is necessary to serve legitimate biosecurity goals.
8. When the government does not fund the research in question, the First Amendment imposes strong limits on the government's ability to restrict the communication of research results, including research that could be used for bioterrorism. When the government funds the research in question, the First Amendment gives it more leeway to restrict the communication of research results, but even in that context, the government's authority may be constrained.
9. Currently, no international organization is giving systematic attention to developing policy or guidance regarding the dissemination of scientific information of concern. Potential mechanisms and institutions [e.g., the World Health Organization (WHO), the Biological Weapons Convention (BWC), the Australia Group (AG), United Nations Security Council (UNSC), etc.] exist that could fulfill this function. There has been a recent decline in policy activity at the international level despite the fact that there are ongoing concerns and discussions about specific technologies (e.g., CRISPR-Cas9).
10. Export controls do not limit communications among U.S. citizens within the United States. Export controls thus have a limited reach and do not offer a mechanism, in and of themselves, to control the dissemination of information.

MECHANISMS AND PROCESSES

A key issue identified during the committee's public meetings and private discussions was how to provide researchers—and particularly journal editors—with guidance about potentially problematic research findings or manuscripts. DURC policies provide mechanisms to guide those carrying out federally funded research or working at institutions that receive federal funds, including requirements to develop, in appropriate cases, risk mitigation plans. Other researchers and journal editors do not have ready access to such guidance. In light of the increasing number of journals in many parts of the world and the utilization of pre-print servers and other means of online publication prior to (or in lieu of) traditional peer review, the situation is significantly more complicated. The following findings relate to U.S. researchers and their international collaborators.

Findings

11. There is no systematic process through which journal editors and researchers outside federally funded institutions can seek guidance from U.S. government experts on the management of manuscripts or on research activities that raise potential biosecurity concerns.
12. There is no shared, consistent policy among U.S. and international journals for addressing DURC.
13. There are limited mechanisms [e.g., the National Science Advisory Board for Biosecurity (NSABB), Federal Bureau of Investigation (FBI) Weapons of Mass Destruction Directorate coordinators] for ongoing engagement between the scientific community and the national security and intelligence communities on biosecurity issues.
14. As a federal advisory body, the NSABB does not have the legal authority to restrict the dissemination of information. The NSABB may provide advice regarding the publication of information only under narrowly defined circumstances. Moreover, knowledge and use of the NSABB throughout the research community is limited.
15. In contrast to the Recombinant DNA Advisory Committee process, the oversight of DURC does not include mechanisms for assessing and sharing of best practices in the management of biosecurity among research institutions or opportunities for high-level review and consultation.
16. In principle, the NSABB could provide a mechanism to fulfill many of the functions described above, but its current mandate is limited.

EDUCATION AND TRAINING

Reaching consensus on the management of DURC is complicated by the fact that experts have fundamentally divergent views about the nature of the biosecurity threat.[2] Any effort to place controls on information for biosecurity purposes involves a careful consideration of the nature of the research, the risks of malevolent uses of the research results, the benefits for scientific advance or the development of countermeasures through open communication, and evaluation of means to reap the benefits while limiting the risks. Effective assessment relies on an appropriate knowledge of risk and policy options among the international community of researchers, funders, and publishers.

Findings

17. Despite the attention given to periodic controversies over DURC, the available evidence suggests that most life scientists have little awareness of issues related to biosecurity. Those training to become life scientists are rarely introduced to the topic in a systematic way. Education and training programs at the undergraduate, graduate, and postdoctorate levels generally do not include courses or discussions about dual use research or DURC, unless the student or trainee is involved in research with a select agent. Even in this case, biosafety is the primary focus. This situation hampers efforts to implement policies to address potential biosecurity risks, particularly in emerging research fields that may pose concerns.
18. The management of the dissemination of scientific information requires local, national, and international approaches to provide awareness-raising, education and training, and ongoing guidance and opportunities to share best practices and develop common approaches.
19. There are some extensive and effective programs at research institutions that deal with specific pathogens that ensure that researchers are trained in biosafety, but they are not systematically in place across U.S. research institutions. In a number of cases, the scope of these programs includes biosecurity and enables these particular communities to develop sophisticated views about these issues. Expanding these programs beyond a focus solely on specific pathogens could increase the ability of the broader research community to take greater responsibility for safeguarding dangerous information in ways that do not impede scientific advances.
20. Lessons learned from experiences with efforts to manage the dissemi-

---

[2] See, e.g., C. Boddie et al., "Assessing the Bioweapons Threat," *Science*, August 21, 2015, Vol. 349, No. 6250, pp. 792-793.

nation of research information are not being adequately assessed or shared so as to promote more effective practice.
21. Many investments have been made by major donors to assist foreign countries with enhancements to their biosafety capacity. Investments have also been made in some aspects of biosecurity (e.g., physical security, access controls, pathogen accounting, etc.). Far fewer resources have been devoted to awareness-raising, education and training, and policy development related to the conduct of research and the dissemination of scientific information that could be employed for bioterrorism.

CONCLUSION

Despite decades of effort, there is little national or international consensus with regard to appropriate policies for addressing issues associated with the conduct and dissemination of life sciences research that might qualify as DURC. The absence of an international commitment to addressing such issues; the lack of agreement regarding a framework for assessing risk, uncertainty, and benefit; and the difficulties the U.S. government has faced in developing policies that effectively manage DURC illustrate the challenges of resolving the issues concerning information dissemination raised by DURC.

Appendixes

Appendix A

Biographical Information of Committee and Staff

CO-CHAIRS

RICHARD A. MESERVE (NAE), J.D., Harvard Law School; Ph.D. (Applied Physics) Stanford University; B.A., Tufts University, is President Emeritus of the Carnegie Institution for Science. Before assuming the Carnegie presidency in April 2003, he was Chairman of the U.S. Nuclear Regulatory Commission (NRC), having served since October 1999. Before joining the NRC, Dr. Meserve was a partner in the law firm of Covington & Burling LLP, where he now serves on a part-time basis as a Senior Of Counsel. He devoted his legal practice to technical issues arising in environmental and toxic tort litigation, counseling scientific societies and high-tech companies, and nuclear licensing. Early in his career, he served as legal counsel to the President's science advisor, and was a law clerk to Justice Harry A. Blackmun of the U.S. Supreme Court and to Judge Benjamin Kaplan of the Massachusetts Supreme Judicial Court. He is a member of the National Academy of Engineering and the American Philosophical Society; a Fellow of the American Academy of Arts and Sciences, the American Association for the Advancement of Sciences, and the American Physical Society; and a Foreign Member of the Russian Academy of Sciences. He currently serves as Chairman of the International Nuclear Safety Group, chartered by the International Atomic Energy Agency, and Co-Chairman of the Department of Energy's Nuclear Energy Advisory Committee. He was formerly President of the Board of Overseers of Harvard University and now serves as a member of the Council of the National Academy of Engineering and of the American Academy of Arts and Sciences. He has previously served on numerous committees and boards of the National Academies of Sciences, Engineering, and Medicine, including as co-chair of the committee on Science, Technology, and Law. Dr. Meserve also serves on the boards of PG&E Corporation and TriAlpha Energy Corporation. He wrote the amicus briefs on behalf of the National Academy of Engineering in the *Kumho* case and on behalf of the National Academy of Sciences in the *Daubert* case. These landmark cases established the basis for admitting expert testimony into court.

HAROLD E. VARMUS (NAS/NAM), M.D., co-recipient of a Nobel Prize in 1989 for studies of the genetic basis of cancer, joined the Meyer Cancer Center of Weill Cornell Medical College as the Lewis Thomas University Professor on April 1, 2015, when he also became a Senior Associate Member of the New York Genome Center. Previously, Dr. Varmus was the Director of the National Cancer Institute (2010-2015), President of Memorial Sloan-Kettering Cancer Center (2000-2010), and Director of the National Institutes of Health (1993-1999). A graduate of Amherst College and Harvard University in English literature and of Columbia University in medicine, he trained at Columbia University Medical Center, the National Institutes of Health, and the University of California San Francisco (UCSF) before joining the basic science faculty at UCSF, where he worked for over two decades (1971-1993). The author of more than 350 scientific papers and 5 books, including a memoir titled *The Art and Politics of Science* (2009), he was a co-chair of President Obama's Council of Advisors on Science and Technology, co-founder and Chairman of the Board of the Public Library of Science, and chair of the Scientific Board of the Gates Foundation Grand Challenges in Global Health. He is a member of the National Academies of Sciences and Medicine and a foreign member of the Royal Society and is involved in several initiatives to promote science and health in developing countries.

MEMBERS

ARTURO CASADEVALL (NAM) is Professor and Chair in the W. Harry Feinstone Department of Molecular Microbiology and Immunology at The Johns Hopkins Bloomberg School of Public Health. Formerly, he was Leo and Julia Forchheimer Professor of Microbiology and Immunology; Chair, Department of Microbiology and Immunology; and Professor, Department of Medicine at the Albert Einstein College of Medicine. He has published more than 700 scientific papers and has co-authored a book on *Cryptococcus neoformans*. He is a Fellow of the American Academy of Microbiology and was elected to the American Society for Clinical Investigation, to the American Association of Physicians, and as a Fellow of the American Association for the Advancement of Science. Dr. Casadevall has served on numerous advisory committees to the National Institutes of Health including study sections, strategic planning for the National Institute of Allergy and Infectious Diseases (NIAID), and the blue ribbon panel on response to bioterrorism. He currently co-chairs the Board of Scientific Counselors for the NIAID and is a former member of the National Science Advisory Board for Biosecurity (NSABB). He is the founding editor of the first American Society for Microbiology general journal, *mBio,* and serves on the editorial boards of several journals, and has been the recipient of numerous awards, most recently the Solomon A. Berson Medical Alumni Achievement Award in Basic Science-New York University School of Medicine, the Infectious Diseases Society of America Kass Lecturer, and the ASM William Hinton Research Training Center Award for mentoring scientists from underrepresented groups.

DENISE CHRYSLER, J.D., is director of the Mid-States Region of the Network for Public Health Law, located at the University of Michigan School of Public Health. The network assists public health practitioners to use law to improve the health of communities. She serves her local community as a member of the Ingham County (Michigan) Board of Health.

For 27 years, Ms. Chrysler provided legal services to Michigan's state health department regarding communicable disease, immunization, environmental public health, public health research, privacy, health information exchange, and emergency legal preparedness and response. She worked extensively on the Michigan BioTrust for Health to make newborn screening blood specimens and associated data available for health research. She served as the state health department's public health legal director, privacy officer, freedom of information coordinator, regulatory affairs officer, and member of the institutional review board. She also represented the health department as an assistant attorney general.

ANUJ C. DESAI is the William Voss-Bascom Professor of Law at the University of Wisconsin, where he teaches in both the Law School and the School of Library and Information Studies, offering classes in copyright, the First Amendment, legislation, legislation and regulation, and cyberlaw. He is currently on leave, serving as an administrative appellate judge, as a member of the Administrative Review Board of the U.S. Department of Labor. He also serves as a part-time Commissioner of the Foreign Claims Settlement Commission, an independent, quasi-judicial agency of the U.S. Department of Justice that adjudicates claims of U.S. nationals against foreign governments.

MICHAEL ETTENBERG (NAE) is Managing Partner at DOLCE Technologies, a company that commercializes technologies invented at leading universities, such as Princeton and Columbia. Previously, he retired from Sarnoff (formerly RCA) Labs after 35 years, ending as Senior Vice President in charge of all of Sarnoff's device research, including small silicon integrated circuit fabrication, TV displays, optoelectronics, and cameras. Dr. Ettenberg has extensive experience with III-V materials and optoelectronic devices. He developed the dielectric mirrors used on all of today's laser diodes. Dr. Ettenberg has published 110 papers and has been awarded 35 patents, mainly in the area of optoelectronics. He also was president of the IEEE Lasers and Electro-Optics Society and a member of the Defense Science Board.

DAVID FIDLER is the James Louis Calamaras Professor of Law at the Indiana University Maurer School of Law and is one of the world's leading experts on international law and global health. His books in this area include *Biosecurity in the Global Age: Biological Weapons, Public Health, and the Rule of Law* (Stanford University Press, 2008) (with Lawrence O. Gostin), *SARS, Governance, and the Globalization of Disease* (Palgrave Macmillan, 2004), *International Law and Public Health: Materials on and Analysis of Global Health Jurisprudence* (Transnational Publishers, 2000), and *International Law and Infectious Diseases* (Clarendon Press, 1999). He has published mote than 100 articles and chapters on global health topics in legal, public health, medical, and political science journals and books.

In addition to his teaching and scholarly activities, Professor Fidler has served as an international legal consultant to the World Health Organization and the U.S. Centers for Disease Control and Prevention. He has twice been appointed by the Director-General of the World Health Organization as a member of the IHR Roster of Experts, the members of which advise the Director-General on matters relating to the International Health Regulations (2005). He is an Associate Fellow with the Centre on Global Health Security at the Royal Institute of International Affairs (Chatham House).

Professor Fidler also specializes in other topics, including international law relating to cybersecurity and cyberspace. He is an Adjunct Senior Fellow for Cybersecurity with the Council on Foreign Relations and is the editor of *The Snowden Reader* (Indiana University Press, 2015).

CLAIRE FRASER (NAM) is Director of the Institute for Genome Sciences and a Professor of Medicine at the University of Maryland School of Medicine in Baltimore, Maryland. She was previously the President and Director of The Institute for Genomic Research in Rockville, Maryland. Dr. Fraser has played a seminal role in the sequencing and analysis of human, animal, plant, and microbial genomes to better understand the role that genes play in development, evolution, physiology, and disease. Her current research interests are focused on the structure and function of the human git microbiota. Dr. Fraser has more than 240 scientific publications and has served on committees of the National Science Foundation, Department of Energy, and National Institutes of Health. She is the recipient of numerous awards and honors including the Promega Biotechnology Award and the E.O. Lawrence Award from the Department of Energy, she is a Fellow of the American Association for the Advancement of Science and the American Association of Microbiology, and she has been elected into the Maryland Women's Hall of Fame and the National Academy of Medicine.

MICHAEL HOPMEIER is the President, Unconventional Concepts, Inc. and has been a technical advisor and operational consultant to numerous governmental agencies including the Defense Advanced Research Projects Agency Defense Sciences Office, U.S. Army Medical Research and Materiel Command, U.S. Surgeon General, and the Deputy Assistant to the Secretary of Defense for Chemical and Biological Defense. He was one of the primary developers of the Bioterrorism Preparedness Program at the U.S. Center for Disease Control and Prevention, served as the Science and Technology Advisor to the U.S. Air Force Surgeon General, as well as the first S&T Advisor to the U.S. Marine Corps Chem/Bio Incident Response Force.

Mr. Hopmeier has been a member and/or task force chair for numerous senior advisory panels including the Defense Science Board and the National Academy of Sciences and served on the Senior Policy and Strategy Panel for Lawrence Livermore National Laboratory. He is a founding member and current member of the Executive Board of the International Counter-Terrorism Academic Community and an Associate Researcher of the Institute for Counter-Terrorism.

Mr. Hopmeier is an internationally recognized expert on countering suicide terrorism, disaster/crisis response. and emergency management and preparedness. He is a founder of a number of different start-up companies and sits on the board of several high-technology firms. He has been involved in numerous international programs as a manager or advisor and has supported a number of efforts in the UK, Greece, and Israel and has authored numerous papers and presentations on topics ranging from biological model development and biotechnology research to emergency response training and suicide bombing.

Mr. Hopmeier's project areas include training and preparedness, chemical/biological incident response, combat casualty care and medical support, crisis response and management, unconventional pathogen countermeasure programs, federal agency protective measures, counter-terrorism, and integrated federal/civilian disaster response.

JAMES LE DUC is Professor in the Department of Microbiology and Immunology, School of Medicine and Director of the Galveston National Laboratory at the University of Texas Medical Branch (UTMB) in Galveston, Texas, where he holds the John Sealy Distinguished University Chair in Tropical and Emerging Virology. The Galveston National Laboratory is a biocontainment facility involved in basic and applied research into highly pathogenic infectious diseases, including Ebola virus. Work under way includes investigations into the development of vaccines, therapeutics and diagnostics for emerging infectious diseases, and agents of potential use in bioterrorism. Prior to joining UTMB, Dr. Le Duc worked at the U.S. Centers for Disease Control and Prevention where he served in various leadership roles, including director of the division of viral and rickettsial diseases, associate director for global health, and coordinator for pandemic influenza preparedness. He had a 23-year career as an Officer in the U.S. Army Medical Research and Development Command with assignments at the Walter Reed Army Institute of Research, the U.S. Army Medical Research Institute of Infectious Diseases, and several overseas duty stations. Dr. Le Duc is an expert in public health, specifically in infectious diseases caused by viruses. Dr. Le Duc is a Fellow of the Infectious Diseases Society of America, a current member of the National Science Advisory Board for Biosecurity, and a member of several other professional societies. Dr. Le Duc was a medical officer in communicable diseases at the World Health Organization (WHO) from 1992 to 1996, was instrumental in implementing the WHO program in emerging infectious diseases, and currently serves as a member of the WHO Global Outbreak Alert and Response Network steering committee. He is a National Associate of the National Research Council.

W. IAN LIPKIN, M.D., the John Snow Professor of Epidemiology and Professor of Neurology and Pathology at Columbia University, is internationally recognized for the development of genetic methods for microbial surveillance and discovery. Dr. Lipkin directs the Center for Infection and Immunity at Columbia University and the National Institutes of Health (NIH) Center for Research in Diagnostics and Discovery, is a member of the Advisory Committee to the Director of the NIH, and Scientific Director of the Joint Research Laboratory for Pathogen Discovery in the Chinese Centers for Disease Control.

A graduate of the University of Chicago Laboratory School and Sarah Lawrence College, Dr. Lipkin obtained his M.D. at Rush Medical College, Medicine Residency at the University of Washington, Neurology, Residency at the University of California San Francisco, and Fellowship in Microbiology and Neuroscience at The Scripps Research Institute in La Jolla, California. His contributions include the first use of genetic methods to identify an infectious agent; implication of West Nile virus as the cause of the encephalitis in North America in 1999; invention of MassTag PCR and the first panmicrobial microarray; first use of deep sequencing in pathogen discovery; and molecular characterization of more than 900 viruses.

At the height of the 2003 severe acute respiratory syndrome (SARS) outbreak, Dr. Lipkin traveled to the People's Republic of China at the invitation of the World Health Organization, the Chinese Minister of Science and Technology, Xu Guanhua, and the Vice President of the Chinese Academy of Sciences (CAS), Chen Zhu, hand-carrying 10,000 test kits to Beijing. After training clinical microbiologists in their use, he returned to New York, became ill, and was placed into quarantine. He nonetheless continued to co-direct SARS research efforts within CAS as Special Advisor through 2004. More recently, he was the sole external investigator to be invited by the Ministry of Health in Saudi Arabia to assist in identifying reservoirs and vectors for transmission of the Middle East respiratory syndrome coronavirus.

Dr. Lipkin has been active in translating science to the public through print and digital media. He acted as chief scientific consultant for the Hollywood film *Contagion*, has been featured in dozens of news publications including *The New York Times*, BBC, and the *Wall Street Journal*. He has appeared on CNN, CBS, ABC, *Nova*, Charlie Rose, and *Through the Wormhole* with Morgan Freeman. In 2012, Dr. Lipkin was named "the world's most celebrated virus hunter" by *Discover Magazine*. His honors include the following: Pew Scholar in the Biomedical Sciences, Japanese Human Science Foundation Visiting Professor, Columbia College of Physicians and Surgeons Visiting Bruenn Professor, American Society of Microbiology Foundation Lecturer, Ellison Medical Foundation Senior Scholar in Global Infectious Disease, Fellow of the New York Academy of Sciences, Distinguished Lecturer of the National Center for Infectious Diseases, Fellow of the American Society for Microbiology, John Courage Professor National University of Singapore, Kinyoun Lecturer National Institutes of Health, Fellow of the Wildlife Conservation Society, Fellow of the American Association for the Advancement of Science, member of the Association of American Physicians, Oxford University Simonyi Lecturer, and recipient of the Villanova University Mendel Medal. In 2016 he received the International Science and Technology Cooperation Award, the top science honor in China for his contributions to the advancement of science in the country.

STEPHEN S. MORSE is Professor of Epidemiology and Director, Infectious Disease Epidemiology Certificate Program, Columbia University Medical Center Mailman School of Public Health. Dr. Morse's interests focus on epidemiology and risk assessment of infectious diseases (particularly emerging infections, including influenza), and improving disease early warning systems.

In 2000, he returned to Columbia after 4 years in government as program manager for biodefense at the Defense Advanced Research Projects Agency, where he co-directed the Pathogen Countermeasures program and subsequently directed the Advanced Diagnostics program. Before joining Columbia, he was assistant professor of virology at The Rockefeller University in New York and remains an adjunct faculty member. His book, *Emerging Viruses* (Oxford University Press) was selected by "American Scientist" for its list of "100 Top Science Books of the 20th Century." Dr. Morse was chair and principal organizer of the 1989 National Institute of Allergy and Infectious Diseases/National Institutes of Health Conference on Emerging Viruses, for which he originated the term and concept of emerging viruses/infections; served as a member of the Institute of Medicine/National Academy of Sciences' Committee on Emerging Microbial Threats to Health (and chaired its Task Force on Viruses) and was a contributor to its report, *Emerging Infections* (1992). He subsequently served on the Steering Committee of the Institute of Medicine's Forum on Microbial Threats and the National Academy of Sciences' committees on biowarfare threats, and as an adviser to numerous government and international organizations. He was the founding chair of ProMED (the nonprofit international Program to Monitor Emerging Diseases) and was an originator of ProMED-mail, an international network inaugurated by ProMED in 1994 for outbreak reporting and disease monitoring using the Internet. Dr. Morse is a current member of the National Science Advisory Board for Biosecurity.

STAFF

ANNE-MARIE MAZZA, Ph.D., is the senior director of the Committee on Science, Technology, and Law. Dr. Mazza joined the National Academies of Sciences, Engineering, and Medicine in 1995. In 1999 she was named the first director of the Committee on Science, Technology, and Law. Dr. Mazza has been the study director on numerous Academy reports including *Optimizing the Nation's Investment in Academic Research* (2016); *International Summit on Human Gene Editing: A Global Discussion* (2015); *Identifying the Culprit: Assessing Eyewitness Identification* (2014); *Positioning Synthetic Biology to Meet the Challenges of the 21st Century* (2013); *Reference Manual on Scientific Evidence*, 3rd Edition (2011); *Review of the Scientific Approaches Used During the FBI's Investigation of the 2001 Anthrax Letters* (2011); *Managing University Intellectual Property in the Public Interest* (2010); *Strengthening Forensic Science in the United States: A Path Forward* (2009); *Science and Security in a Post 9/11 World* (2007); *Reaping the Benefits of Genomic and Proteomic Research: Intellectual Property Rights, Innovation, and Public Health* (2005); and *Intentional Human Dosing Studies for EPA Regulatory Purposes: Scientific and Ethical Issues* (2004). Between October 1999 and October 2000, Dr. Mazza divided her time between the National Academies and the White House Office of Science and Technology Policy, where she served as a senior policy analyst responsible for issues associated with a Presidential Review Directive on the government-university research partnership. Before joining the National Academies, Dr. Mazza was a senior consultant with Resource Planning Corporation. She is a fellow of the American Association for the Advancement of Science. Dr. Mazza was awarded a B.A., M.A., and Ph.D. from George Washington University.

JO L. HUSBANDS, Ph.D., is a Scholar/Senior Project Director with the Board on Life Sciences, where she manages studies and projects to help mitigate the risks of the misuse of scientific research for biological weapons or bioterrorism. She represents the National Academies of Sciences, Engineering, and Medicine on the Biosecurity Working Group of IAP: The Global Network of Science Academies, which also includes the academies of Australia, China, Cuba, Egypt, India, Nigeria, Poland (chair), and the United Kingdom. From 1991 to 2005 she was Director of the National Academies' Committee on International Security and Arms Control and its Working Group on Biological Weapons

Control. Before joining the National Academies, she worked for several Washington, DC-based nongovernmental organizations focused on international security. Dr. Husbands is currently an adjunct professor in the Security Studies Program at Georgetown University, where she teaches a course on the International Arms Trade. She is a member of the Honor Roll of Women in International Security, the International Institute for Strategic Studies, and the Global Agenda Council on Nuclear, Chemical, and Biological Weapons of the World Economic Forum. She is also a fellow of the International Union of Pure and Applied Chemistry. She holds a Ph.D. in political science from the University of Minnesota and a master's degree in international public policy (international economics) from The Johns Hopkins University School of Advanced International Studies.

STEVEN KENDALL, Ph.D., is program officer for the Committee on Science, Technology, and Law. Dr. Kendall has contributed to numerous National Academies of Sciences, Engineering, and Medicine reports, including *Optimizing the Nation's Investment in Academic Research* (2016); *International Summit on Human Gene Editing: A Global Discussion* (2015); *Identifying the Culprit: Assessing Eyewitness Identification* (2014); *Positioning Synthetic Biology to Meet the Challenges of the 21st Century* (2013); the *Reference Manual on Scientific Evidence*, 3rd Edition (2011); *Review of the Scientific Approaches Used During the FBI's Investigation of the 2001 Anthrax Mailings* (2011); *Managing University Intellectual Property in the Public Interest* (2010); and *Strengthening Forensic Science in the United States: A Path Forward* (2009). Dr. Kendall completed his Ph.D. in the Department of the History of Art and Architecture at the University of California, Santa Barbara, where he wrote a dissertation on 19th century British painting. Dr. Kendall received his M.A. in Victorian art and architecture at the University of London. Prior to joining the National Academies of Sciences, Engineering, and Medicine in 2007, he worked at the Smithsonian American Art Museum and The Huntington in San Marino, California.

KAROLINA KONARZEWSKA is program coordinator for the Committee on Science, Technology, and Law. She is a master's student of economics at George Mason University. She holds a master's degree in international relations from New York University and a bachelor's degree in political science from the College of Staten Island, City University of New York. Prior to joining the National Academies of Sciences, Engineering, and Medicine, Ms. Konarzewska worked at various research institutions in Washington, D.C., where she covered political and economic issues pertaining to Europe, Russia, and Eurasia.

KARIN MATCHETT, Ph.D., is a freelance writing consultant who works on topics in science, technology, and environment. Her work spans all phases of a document's development—from sharp outline to first draft to polished product.

Dr. Matchett has done developmental evaluations and substantive editing for more than 200 research grants in academic settings and written strategic visioning documents, summaries of expert panels in academia, and proposals for academic program development and research. She also works with nonprofit organizations to develop reports, proposals, and web content.

Dr. Matchett has a Ph.D. in the history of science from the University of Minnesota, with an emphasis on 20th century life sciences and agriculture in the United States and Mexico. She completed a post-doctoral fellowship under the mentorship of Daniel Kevles at Yale University in which she did research and writing on topics at the intersection of the life sciences and law. Her current research focus is in energy and climate issues as they relate to human psychology, world history, and American society and culture.

Appendix B

Committee Meeting Agendas

MEETING 1
New York, NY
JULY 11-12, 2016

MONDAY, JULY 11, 2016

OPEN SESSION

10:00 am   Welcome and Introductions

                Committee Co-chairs:

                Richard A. Meserve, Covington & Burling LLP
                Harold E. Varmus, Weill Cornell Medicine

10:15 am   Charge from Sponsors

                Paula Olsiewski, Alfred P. Sloan Foundation
                Ed You, Federal Bureau of Investigation

10:30 am   Overview of Government Policies Influencing Publication of Dual Use Research of Concern

                Speaker:

                Gerald L. Epstein, White House Office of Science and Technology Policy

10:50 am	Key Challenges of Current Policy and Possible Options for Limited Dissemination
	Speaker:
	Elisa D. Harris, University of Maryland
11:15 am	Committee Discussion with Drs. Epstein and Harris
12:00 pm	Lunch
1:00 pm	Keeping Up with Dual Use Research, Emerging Science, and Publication Concerns: Challenges for Scientific Journals
	Speakers:
	Philip Campbell, *Nature*—via videoconference
Inder Verma, *Proceedings of the National Academy of Sciences of the United States of America*—via videoconference
Randy Schekman, *eLife*—via videoconference
2:30 pm	Looking Forward: Lessons Learned from the Past and Options for the Future
	Speakers:
	Michael Imperiale, University of Michigan
David A. Relman, Stanford University and VA Palo Alto Medical Center
4:00 pm	Break
4:15 pm	National Science Advisory Board for Biosecurity's Current Thinking on Publication of Dual Use Research of Concern
	Speaker:
	Carrie Wolinetz, National Institutes of Health
5:00 pm	Adjourn to Closed Session

TUESDAY, JULY 12, 2016

OPEN SESSION

8:30 am	Breakfast
9:00 am	Welcome

                Committee Co-chairs:

                Richard A. Meserve, Covington & Burling LLP
                Harold E. Varmus, Weill Cornell Medicine

9:15 am   Lessons Learned

                Speaker:

                Teresa Hauguel, National Institute of Allergy and Infectious Diseases

10:00 am   Break

10:15 am   Perspectives from Research Institutions

                Speakers:

                Ara Tahmassian, Harvard University
                David L. Wynes, Emory University

11:15 am   Considerations for Options for Limiting/Restricting Dissemination

                Speaker:

                Alan B. Morrison, The George Washington University—via videoconference

12:00 pm   Lunch

1:00 pm   Adjourn to Closed Session

MEETING 2
Washington, DC
JANUARY 4, 2017

WEDNESDAY, JANUARY 4, 2017

OPEN SESSION

8:30 am	Breakfast
9:00 am	Welcome and Opening Remarks
	Committee Co-chairs:
	Harold E. Varmus, Weill Cornell Medicine Richard A. Meserve, Covington & Burling LLP
9:15 am	Dual Use Research of Concern—The Biosafety/Biosecurity Context
	Speaker:
	Joseph A. Kanabrocki, The University of Chicago
9:30 am	Discussion
10:00 am	Mechanisms for Managing Dual Use Research of Concern I
	Moderator:
	Nancy Connell, Rutgers, The State University of New Jersey
	Speakers:
	David A. Relman, Stanford University and VA Palo Alto Health Care System and Michael Imperiale, University of Michigan—via videoconference Sam Weiss Evans, Harvard University
10:30 am	Discussion
11:15 am	Break

APPENDIX B

11:30 am	Mechanisms for Managing Dual Use Research of Concern II
	Moderator:
	Gigi Kwik Gronvall, The John Hopkins Bloomberg School of Public Health
	Speakers:
	Tim Stearns, Stanford University Duane Lindner, Sandia National Laboratories
12:00 pm	Discussion
12:45 pm	Lunch
2:00 pm	Mechanisms for Managing Dual Use Research of Concern III
	Moderator:
	Margaret E. Kosal, Georgia Institute of Technology
	Speakers:
	Piers D. Millett, Biosecure Ltd. Kimberley Strosnider and Doron Hindin, Covington & Burling LLP
2:30 pm	Discussion
3:15 pm	Open Session Adjourns

Appendix C

Acronyms and Abbreviations

AAAS	American Association for the Advancement of Science
AG	Australia Group
ASM	American Society for Microbiology
BWC	Biological Weapons Convention
CRISPR	Clustered regularly interspaced short palindromic repeats
CUI	Controlled unclassified information
DTRA	Defense Threat Reduction Agency
DURC	Dual use research of concern
EAR	Export Administration Regulations
EASAC	European Academies Scientific Advisory Council
FBI	Federal Bureau of Investigation
FESAP	Federal Experts Security Advisory Panel
FOIA	Freedom of Information Act
GOF	Gain-of-function
GOFROC	Gain-of-function Research of Concern
HHS	U.S. Department of Health and Human Services
HPAI	Highly Pathogenic Avian Influenza
ICSU	International Council for Science
iGEM	international Genetically Engineered Machines competition
IRE	Institutional Review Entity
ITAR	International Traffic in Arms Regulation

MERS	Middle East respiratory syndrome
NATO	North Atlantic Treaty Organization
NIH	U.S. National Institutes of Health
NRC	National Research Council
NSABB	National Science Advisory Board for Biosecurity
NSDD-189	National Security Decision Directive-189
OSTP	White House Office of Science and Technology Policy
PATRIOT Act	Uniting and Strengthening America by Providing Appropriate Tools Required to Intercept and Obstruct Terrorism Act of 2001
PI	Principal Investigator
PIP	Pandemic Influenza Preparedness Framework
PNAS	*Proceedings of the National Academy of Sciences of the United States of America*
PPP	Pathogens of pandemic potential
SARS	Severe acute respiratory syndrome
Synbio LEAP	Synthetic Biology Leadership Excellence Accelerator Program
UN	United Nations
UNESCO	United Nations Educational, Scientific and Cultural Organization
UNSC	United Nations Security Council
WHO	World Health Organization